Springer Proceedings in Business and Economics

More information about this series at http://www.springer.com/series/11960

Jorge Marx Gómez • Brenda Scholtz
Editors

Information Technology in Environmental Engineering

Proceedings of the 7th International Conference on Information Technologies in Environmental Engineering (ITEE 2015)

Editors
Jorge Marx Gómez
Department of Computing Science
University of Oldenburg
Oldenburg, Germany

Brenda Scholtz
Department of Computing Science
Nelson Mandela Metropolitan University (NMMU)
Port Elizabeth, South Africa

Springer Proceedings in Business and Economics
ISBN 978-3-319-79742-7 ISBN 978-3-319-25153-0 (eBook)
DOI 10.1007/978-3-319-25153-0

Springer Cham Heidelberg New York Dordrecht London

Softcover reprint of the hardcover 1st edition 2016

Printed on acid-free paper

Springer International Publishing AG Switzerland is part of Springer Science+Business Media (www.springer.com)

Preface

Since 2003 the conference series Information Technology in Environmental Engineering (ITEE) has established a platform for discussing the progress in the field. During the 6th ITEE 2013 conference, practitioners and scientists met at the Leuphana University Lüneburg (Germany) to discuss recent developments, promising ideas, and new challenges in information management for supporting sustainability efforts. The 7th ITEE in 2015 was hosted for the first time outside of Europe in Port Elizabeth, South Africa. The conference aimed at highlighting the impact companies and other large organisations have on limited global resources and how this can be improved through the use of technology. Setting up smart home and organisational environments to efficiently manage the consumption of scarce resources such as water and electricity will be one of the themes of the conference. Sustainability and integrated reporting is another key issue that will be addressed. We were very pleased to welcome both international and national researchers and professional representatives from industry to the conference.

The Conference Co-Chairs of ITEE 2015 and the editors of this book would like to extend their gratitude to all those who made the conference a success and contributed to this book. We would also like to thank all authors and participants of the conference for their contributions. We greatly appreciate the commitment and support of the programme committee, namely Kevin Naudé, Lester Cowley, Andre Calitz, Jean Greyling, Kerryn Botha, Bianca Deyzel, Mareike Hinrichs, Jean Rademakers, and Hayley Irvine. Special thanks must be extended to the assistant editors Clayton Burger, Samantha Ludick, and Maxine Esterhuyse for their assistance with checking of articles and editing assistance.

Oldenburg, Germany
Port Elizabeth, South Africa

Jorge Marx Gómez
Brenda Scholtz

Contents

Contribution of Mobile Phones to Township Livelihood Outcomes in the Western Cape Province of South Africa

Unathi September, Upendo Fatukubonye, Kevin Allan Johnston, and Brian O'Donovan

1 Introduction

Over the past few years poor communities in developing countries have increasingly adopted mobile phones in their day-to-day activities, and projections suggest that this will continue (Sife et al. 2010). There are many ways in which mobile phones can contribute to livelihoods and quality of life among the poor in developing countries. Mobile phones can quickly and easily transfer information at a relatively low cost, and are not limited by many of the factors that have influenced other communication media (Frempong 2011). Despite this, the relationship between mobile phones, livelihoods and the poor in developing countries has not been sufficiently understood (McNamara 2008; Rashid and Elder 2009), partly due to mixed interpretations of the concepts of sustainable livelihoods and the poor and an inadequate understanding of the nature of mobile phones. Therefore, it is of great interest to investigate how mobile phones contribute to the livelihoods of communities that are deemed to be poor (Goodman 2005).

Even though there is enthusiasm for Information Communication Technologies (ICT's) potential to improve livelihood outcomes, there are still diverging views on how mobile technology contributes to these (Andrade and Urquhart 2009). In addition, a great deal of research has focused on poor communities who have encountered ICTs as a result of a specific development initiative, like a telecentre, rather than mobile phones (Sife et al. 2010).

The purpose of this paper is to explore the contribution of mobile phones to livelihood outcomes amongst the urban poor in South Africa. This was done by replicating a study done by Sife et al. (2010) in Tanzania. The study in this paper was conducted in four Western Cape urban townships. The conditions provide a unique opportunity to extend the understanding of the contribution of mobile

U. September • U. Fatukubonye • K.A. Johnston (✉) • B. O'Donovan
University of Cape Town, Cape Town, South Africa
e-mail: septemberu@gmail.com; upendo.fatukubonye@gmail.com; kevin.johnston@uct.ac.za

J. Marx Gómez, B. Scholtz (eds.), *Information Technology in Environmental Engineering*, Springer Proceedings in Business and Economics,
DOI 10.1007/978-3-319-25153-0_1

phones to the improvement of livelihood outcomes among the urban poor in South Africa. The study will thus be of value to both practitioners and researchers.

This paper is structured as follows: The following section reviews the literature on the poverty in urban areas and summarizes past findings in the light of the Sustainable Livelihoods Framework. The following two sections outline the research methodology adopted, analyze the data and present the findings. The final section summarizes the outcomes of the findings, discusses what lessons can be learnt and offers some recommendations.

2 Literature Review

2.1 The Role of Mobile Phones in Poverty Reduction

South Africa has a long history of racial discrimination and divisions, which resulted in poverty being concentrated in areas such as townships and rural areas. There are a significant number of black South Africans in the townships classified as below the upper bound poverty line (Adato 2007; Maki 2009). The percentage of the black population below the poverty line was 66.8 % in 2006 and 54.0 % in 2011 (Statistics South Africa 2014). Since 1996, poverty eradication has been one of the principal focuses for the South African government. However, poverty still remains a concern not only to South Africa but to other developing countries (Fan 2008). There are conflicting views among the South African Government and other development stakeholders with regards to the definition and measurement of poverty (Frye and Magasela 2005). Poverty can be defined indirectly using a monetary value, or directly by using a set of indicators (Studies in Poverty and Inequality Institute 2007). The direct method is often expressed in terms of poverty lines or thresholds, referring to the income required to avoid poverty, however conceptualised. The direct method uses a set of indicators, which in some instances are combined to form an index (SARPN 2007).

The use of cash income as a poverty indicator in townships and rural areas is limited, due to difficulties in obtaining accurate income data as people in these circumstances rarely record their income (Samuel et al. 2005). However, socio-economic characteristics such as households' income earning activities, the quality of housing of participating township dwellers, and the availability of basic services such as water and electricity can be assessed (Everatt and Smith 2008). These characteristics indicate the degree of poverty of the sample.

The success of ICTs in Western countries has created enthusiasm about their potential for development in African countries. Many have advocated mobile phones as the ideal way to solve socio-economic problems among the poor. This assumption has sparked debates around the way in which mobile phones can contribute to improving livelihood outcomes. The poor are found both in townships and rural areas. A common element in these areas is a lack of affordable access to relevant information and knowledge, considered one of the main contributors to poverty (Bhavnani et al. 2008). ICTs can thus play a major contribution in alleviating poverty (Bhavnani et al. 2008). Studies on the African continent (Mbiti and Weil 2011; Shackleton 2007; Sife et al. 2010) have shown that the use of mobile phones has played a role in assisting poverty eradication initiatives. A framework that is particularly useful for assessing this contribution is the Sustainable Livelihoods Framework (SLF) (Levine 2014).

2.2 *The Sustainable Livelihoods Framework*

The SLF shows that poor communities operate in a context of vulnerability where they have access to different livelihood assets (Fig. 1). Through different transforming structures and processes, livelihood assets gain their value and meaning. The vulnerability context influences livelihood strategies and brings beneficial livelihood outcomes (Levine 2014; Sife et al. 2010). The vulnerability context attempts to describe the extent to which people in that context can withstand the impact of shocks such as fires, burglaries, and loss of employment, trends such as an economic downturn, and seasonality such as a poor crop for a small farmer. However, as Levine (2014) points out, explaining the context of vulnerability is complex and no such context can be fully explained.

2.3 *Livelihood Assets*

Livelihood assets can be said to constitute livelihood building blocks (Farrington et al. 1999). To a limited extent they can be substituted for each other. Thus, the

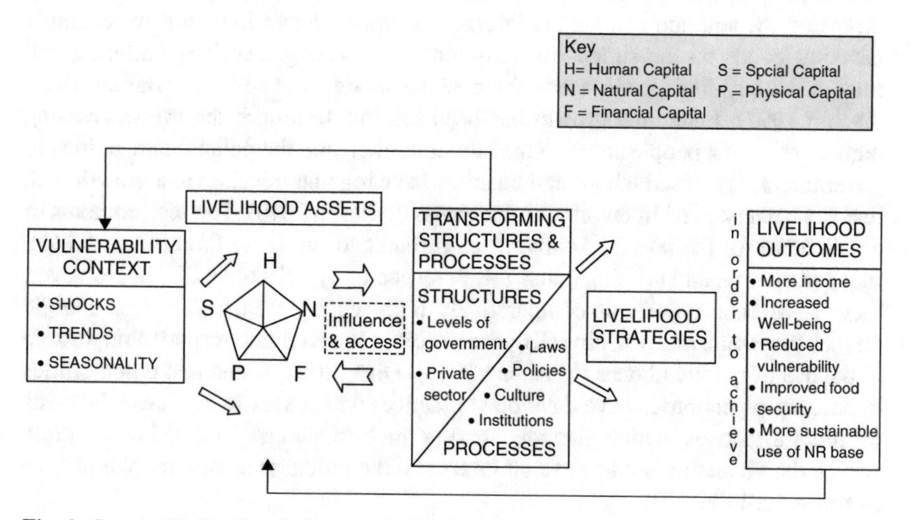

Fig. 1 Sustainable livelihoods framework (Levine 2014)

poor may draw on social capital such as family or neighbourhood security mechanisms at times when financial capital is in short supply (Baumann 2002). There are many variables that may influence mobile phones' adoption and usage, including the concept of the digital divide.

Physical capital refers to those physical aspects that influence mobile use such as the phone itself, and access to electricity and other services. The ownership of mobile phones in township communities is growing at an exceptional rate. In a 2009 study in one of the townships in the Western Cape, 75 % of respondents owned a mobile phone, while the remainder (25 %) said they had used a mobile phone (Kreutzer 2009). Sharing of mobile phones has been common in African countries but more than half report that they do not share their phones (Aker and Mbiti 2010). A mobile phone has become the most used ICT tool (Shackleton 2007) and is inherently suited to remote areas with poor infrastructure (Kefela 2010). MTN has been found to be the most popular network provider in South Africa with 86 % of township respondents using it (Kreutzer 2009). Vodacom, the nation's biggest network, had only 15 % usage among township respondents while Cell C and Virgin Mobile were even more distant (7.6 and 3 %, respectively) (Kreutzer 2009).

The government has initiated programmes to provide electricity, water and basic services to the townships. It is clear that good progress has been made in urban areas although the same cannot be said for rural areas (Everatt and Smith 2008). In a survey of eight urban townships, it was found that 49 % had water in their homes, 44 % had water piped to their yards, 95.9 % had an electricity supply, and 1 % were connected to an unpaid supply (Everatt and Smith 2008).

Financial capital generally refers to monetary resources available to access livelihood options or possibilities (Duncombe 2006). Specifically, in urban townships, it refers to sources of funds such as income from employment or micro businesses, and government grants, and uses of funds such as household expenditure and savings. Nearly half of the people in South African townships are not in full time employment and rely on the informal or microeconomic sector, for example hawking goods on the streets, or non-monetary exchanges such as bartering and trade to satisfy their basic needs for essential goods and services (Barbarin and Khomo 1997). Economic growth has been too low to absorb the ever-increasing number of young people entering the labour market, and the policies and actions of government, organised labour and business have together resulted in a growth path that has been skewed in favour of joblessness (Tshoose 2010). The biggest items of expenditure for the poor in South Africa (urban and rural) are food (about 50 %), fuel and energy, and housing (each 7–8 % respectively) (Kreutzer 2009). However, there have been reports of comparatively high levels of expenditure on mobile phones by people in townships (Kreutzer 2009). Financial factors play a major role in the use of mobile phones (Masiteng 2011). Diga (2007) found that communities in developing countries have developed practices which seek to minimise the costs of mobile phones which include sharing mobile phones and using multiple sim-cards. Financial factors have an impact on the purchase of new mobile phones (Aker and Mbiti 2010).

Human capital refers to the skills, knowledge and ability to work in order to access livelihood options or possibilities. Included in human capital are good health and local formal and informal knowledge (Duncombe 2006). Some have argued for a relationship between mobile phone users and level of education (Samuel et al. 2005), while others maintain that there is no relationship as phone usage only requires functional literacy (Sife et al. 2010). Large parts of the South African population are living in educationally deprived situations (Evoh 2009). This affects the extent to which some of mobile phone features can be used. The low cost of SMSs and the ability to use indigenous languages are factors that have caused an increase in SMS communication. In comparison to the United Kingdom where the ratio of number of outgoing calls to outgoing SMS is 0.6:1, South Africa has a ratio of about 3:1 for prepaid users. For rural communities in South Africa the ratio is 17:1. This highlights the impact literacy levels can have on the use of one of the features of mobile phones (Warren 2006).

Social capital refers to aspects of social organisation that enable access to livelihood options or possibilities (Duncombe 2006). Whilst social networks and social relations are clearly social capital, norms and trust are important in this capital. Social capital can be split into socio-cultural and socio-political capital and natural capital can be omitted as it does not have sufficient relevance for ICT.

2.4 *Transforming Structures and Processes*

Livelihood structure and processes include those aspects of public, private and non-governmental organisations that combine to set policy, deliver services or goods that affect the possibilities for livelihood possibilities. Mobile phone users in townships face technical challenges like weak network signals in certain geographical areas and unreliable electricity supply (Buys et al. 2009). Mobile phone service providers consider installing cell towers in sites where there is a high demand, low installation costs and low maintenance costs. The ease of access to these sites and topography are one of the major determinants of whether a cell tower should be installed or not (Buys et al. 2009). The effects of low income has paved a way for innovative means to reduce costs while at the same time communicating messages effectively and MXit is one such innovation (Diga 2007).

2.5 *Livelihood Outcomes*

Using livelihood assets, people are likely to pursue multiple activities and outcomes. They may, for instance, depend on their own farming, sell their labor locally, or migrate to other areas to find work, all within the same year. Outcomes will neither be simply monetary, nor even tangible in all cases (Baumann 2002).

In terms of financial capital outcomes, mobile phones have contributed towards sending and receiving money in developing countries through the use of M-Pesa, a mobile money service which reached approximately 38 % of Kenya's adult population in 2010 (Jack and Suri 2010). Mobile phones can impact human capital outcomes since they are vital for the success of small and micro businesses (Stillman et al. 2010). The use of mobile phones for business activities in rural areas has enabled business owners to access better markets and directly communicate with customers (Sife et al. 2010). In contrast, in urban areas, micro business owners are more likely to use mobile phones for personal rather than business communication (Chew et al. 2010).

Socio-political capital outcomes have not shown much impact from mobile phone use. Despite its promise there is little evidence of mobile phone owners benefitting from e-governance in developing countries (Reijswoud and Jager 2006). Governments of developing countries continue to lag in providing people with services using mobile phones or internet (West 2006). The non-availability of government information to the public is one of the reasons why there is minimal communication between government and people (Rao 2009).

Mobile phones play a vital role in socio-cultural capital outcomes since they promote maintaining social relationships amongst friends and families (Goodman 2005). Many find it more cost effective to call relatives and friends instead of travelling to see them (Goodman 2005; Kreutzer 2009). Social activities such as burials, weddings, community meetings and religious activities form an important part of the life of communities in developing countries. Mobile phones frequently contribute to the arrangement of these social activities (Sife et al. 2010). Mobile phones are regularly used to initiate some form of inter-personal communication (Donner 2007; Kreutzer 2009). Not surprisingly the number of calls made, and number of text messages sent increases with income (Samuel et al. 2005). Low income users have found clever ways to minimise their call costs. Examples are the use of 'please call me' messages or 'beeping'. Over the past years, the number of mobile phone features other than communication has been rising. Almost all new mobile phones provide the ability to play music, take pictures, record videos, and play games (Tshoose 2010). Several mobile phone manufacturers have developed special phones in recent years that have omitted some of these features in order to provide more affordable phones (Kreutzer 2009). In one study, 12 % of respondents owned one of the most prominent no-frills kinds of phones (Nokia's 1100 and 1600).

2.6 *Identified Gaps and Expected Contribution of the Study*

In spite of the increasing recognition of the potential impact mobile phones could have on improving livelihood outcomes, there is still debate on the ways in this can be achieved among the poor. Many researchers are still concerned about the lack of evidence on how mobile phones are linked to livelihood outcomes in poor areas.

The available evidence on the topic is considered largely subjective and dominated more by promises than reality (Kenny 2002). Few research studies (Goodman 2005; Samuel et al. 2005) have attempted to understand the relationship between mobile phones and livelihood outcomes among the poor in South Africa and even in developing countries (McNamara 2008). Thus this study is important as it adds to the body of knowledge on this subject.

3 Research Method

Using the SLF, this quantitative study explored how mobile phones contribute to township livelihood outcomes and poverty reduction. The framework was used in order to ensure that important aspects of livelihood outcomes were considered. A questionnaire, adapted from the study conducted by Sife et al. (2010) was used to investigate demographic characteristics, socio-economic status, access and usage of mobile phones, the contribution and restrictions of mobile phones to livelihood outcomes. The questionnaire consisted of closed-ended questions with tick boxes and Likert type scales. This study was conducted in four different Western Cape urban townships, namely Gugulethu, Khayelitsha, Mandalay and Langa. These areas were chosen because they have easily accessible community centers and churches. This was important as there were challenges in finding respondents who were willing to answer the full questionnaire because of the number of questions that they had to answer. The sample size for the study was 89 respondents randomly selected in those townships. This will limit the generalisability of the results but will contribute to a better understanding of the use and benefits of mobile phones among the poor.

4 Results and Discussion

The analysis of the results was structured around the SLF commencing with the context of vulnerability. One of the key aspects of the context of vulnerability is demographic characteristics (Levine 2014). In this study, 79.8 % of the respondents were between the ages of 18 and 40. This finding supports other research where it was found that most mobile phone owners come from the 25–45 age groups (Kreutzer 2009; Sife et al 2010). Those between 25 and 45 years of age are regarded as the active cohort with the capability of adopting new technologies such as mobile phones (Sife et al. 2010).

4.1 Livelihood Outcomes

In terms of physical capital, the proportion of households connected to electricity was reported by the respondents to be 98.9 %. This is close to the findings of Everatt and Smith (2008) where 95.9 % had an electricity supply. The proportion of poor households connected to piped water was reported by the respondents to be 86.5 %. In the Everatt and Smith (2008) survey the corresponding figure was 93 %. The number who reported living in houses including rented as well as owned houses was 75.3 %. This statistic ties in with the national census of 2011 where 77.6 % lived in formal houses, 7.9 % in traditional houses and 13.5 % in informal houses (Statistics South Africa 2011).

In the section on financial capital, more than half of the respondents (66 %) reported that they had formal employment, with the majority being domestic workers, security guards and teachers. Twenty seven per cent of the respondents were involved in small or micro businesses such as electric repairs, handy-man, and selling recharge vouchers. Households that depend on grants as their only reliable income source, or depend on small unstable businesses, are regarded as poor. The percentage in full time employment is higher than that reported by Everatt and Smith (2008). Households that are regarded as structurally poor often have no formal work or depend only on one formal job that is insufficient given the household size (Adato et al. 2006). Despite the apparent low levels of income, this study found that the respondents spend on average R35.25 ($3) per week on mobile phones. Compared to the study by Kreutzer (2009), there is a difference of R5.51 which may be due to inflation.

Responses to questions related to human capital revealed that 27 % of the respondents had adult or post-secondary school education, 53 % had attended high school, while 18 % had only attended primary school. These demographics are a characteristic of poor township inhabitants (Krueger and Maleckova 2003; Tilak 2006). Krueger and Maleckova (2003) argue that there is a relationship between the level of education and poverty. People with no education or with only primary or high school level of education are most likely to be poor as it is higher education which provides skills that could be useful in the labour market. Even if primary and high school imparts some valuable attributes, in terms of attitudes and skills they are not sufficient to alleviate poverty (Tilak 2006). It is usually higher education that can take people above the poverty line by increasing the social, occupational and economic mobility of households.

With regards to social capital, all respondents reported that they had previously used a mobile phone to initiate at least one inter-personal communication (not just receiving them), which includes making a phone call or sending an SMS. More than half (54 %) preferred making phone calls to sending SMS. The respondents use mobile phones mostly for personal communication and entertainment. Despite this, 42 % of the respondents had mobile phones with Internet connection and features such as radio, camera and music player. These results contradict Kreutzer's (2009) findings that poor people in developing countries tend to use the 'no-frills' kinds of mobile phones. Several mobile phone manufacturers have developed special mobile phones which have these features and are still sold at a cheaper price. Of the respondents, 52 % felt that the high cost of phones was not a constraint or only a slight constraint.

4.2 *Transforming Structures and Processes*

Respondents were presented with a list of possible constraints to livelihood outcomes and were asked to indicate the degree to which mobile phones had contributed to each. The degree of impact was indicated by a four-point Likert type scale (1 = major constraint, 2 = constraint change, 3 = slight constraint, 4 = not a constraint). Thirty-six per cent of the respondents reported subscribing to more than one network. One reason for this was discussed under financial capital. By a small margin (4 %), MTN was found to be the most popular network with 47 % of respondents using MTN, followed by 44 % using Vodacom. Almost a quarter of the respondents were using Cell C (25 %) and only 1 % using Virgin Mobile. These findings support the findings of Kreutzer (2009), who maintains that MTN is the most popular within townships followed by Vodacom and Cell C. The reason for respondents opting to use MTN may be due to its reliability, as 78 % of respondents reported high network connections with MTN. Most (66 %) did not find their network weak or unreliable. In addition, 71 % had little or no difficulty with a lack of electricity supply.

Most respondents 64 % did not feel that they had any difficulty in getting mobile services, particularly recharge vouchers. This could be because of the increasing number of small businesses that deal in mobiles and sell recharge vouchers in the townships. Some of the respondents (12 %) generated income from selling recharge vouchers. Two-thirds (67 %) of respondents reported that a weak and unreliable network is one of the constraints in townships for mobile phone users. This could be as a result of lack of advanced network infrastructure to support the increasing number of mobile phones in less developed geographical areas. Weak signals in certain geographical areas in developing countries are still influencing mobile phone use (Buys et al. 2009).

4.3 *Livelihood Outcomes and Mobile Phones*

The respondents were presented with a list of possible livelihood outcomes and were asked to indicate the degree to which mobile phones had contributed to each (Table 1). The degree of impact was indicated by a four-point Likert type scale (1 = worsened, 2 = no change, 3 = improved, 4 = greatly improved). A mean score of 2.5 and above denoted that mobile phones have contributed positively towards a

Table 1 Contribution of mobile phones to rural livelihood outcomes

Livelihood and poverty aspects	Poverty and livelihood aspect	N	Worsened	No change	Improved	Greatly improved	Mean
Relationships/contacts with friends/relatives	Sc	88	4.5	11.4	48.9	35.2	3.1
Arranging social functions	Sc	80	5.0	30.0	43.8	21.3	3.1
Contact with members of groups/networks	Sc	88	5.7	25.0	31.8	37.5	3.0
Help in case of emergencies	H	88	5.7	25.0	31.8	37.5	3.0
Improving efficiency of daily activities	H	85	5.9	23.5	47.1	23.5	2.9
Arranging travelling/transport	H	85	5.9	23.5	47.1	23.5	2.9
Improving business	F	67	9.0	55.2	23.9	11.9	2.4
Contributing to household income	F	88	13.6	50.0	29.5	6.8	2.4
Sending/receiving money	F	84	6.0	58.3	33.3	2.4	2.3
Communicating with government depts.	Sp	79	11.4	55.7	24.1	8.9	2.3

S social capital, *Sc* socio cultural, *Sp* socio political, *F* financial capital, *H* human capital

livelihood outcome, whereas a mean score of 2.4 and below denotes a negative contribution.

4.3.1 Financial Capital Outcomes

The respondents indicated a mean score of less than 2.4 for mobile phone contribution towards improving business. Out of 67 respondents who answered the question, more than half (55 %) said mobile phones did not contribute to their businesses, and 9 % said the status of their businesses was worsened. These results are in accordance with the literature (Chew et al. 2010). However, in a rural environment Sife et al. (2010) found that mobile phones have a significant contribution towards businesses. The latter study, conducted in Tanzania, focused more on business owners in the agricultural sector, while in the Western Cape townships most business owners have micro trading businesses. The responses on the contribution of mobile phones to household income showed that 50 % believed that their household income had not been improved by mobile phones, and 13.6 % believed that it had worsened.

Nearly two thirds of respondents (64 %) indicated that the ownership of a mobile phone has not significantly contributed towards sending or receiving money ($\mu = 2.3$). This is in contrast to Jack and Suri (2010) and Ngugi and Pelowski (2010), who suggest that mobile phones have contributed towards sending money through services like M-pesa. One possible reason for this difference could be the lack of awareness of services such as M-pesa, another could be that in South African urban townships banking facilities are more accessible to dwellers. Nearly one quarter (23 %) of respondents indicated that low income is a major constraint to using mobile phones. A further 47 % indicated that low income is at least some constraint to use. Fifteen percent of respondents felt that the high cost of mobile phones is a major constraint on the use of mobile phones and a further 56 % indicated that the high cost is at least some constraint. These results confirm the contention that financial factors strongly influence the use of mobile phones (Masiteng 2011; Warren 2006).

4.3.2 Human Capital Outcomes

Almost four-fifths (79 %) of respondents believed that ownership of a mobile phone has a positive (improved or greatly improved) contribution to receiving support in emergencies ($\mu = 3.0$). The results are not a surprise since Western Cape townships are characterised by high crime rates and are far from social services such as hospitals. These results support the contention that mobile phones contribute positively to assisting in times of emergency (Gough 2005; Sife et al. 2010). Over four-fifths (83 %) of respondents indicated a positive contribution to the efficiency of daily activities ($\mu = 2.9$). These results support Sife et al. (2010), who reported that mobile phones are helpful in enabling township dwellers to perform their daily activities efficiently. This could also have a social capital implication as mobile phones are frequently used for the arrangement of social events (Sife et al. 2010).

Despite the argument that education levels impact negatively on mobile phone use (Evoh 2009), 46 % reported that education levels were not a constraint and a further 20 % reported a slight constraint. Additionally, 71 % felt that a lack of skills placed little or no restriction on the use of a mobile phone. It is a surprise to find that 60 % felt that difficulties with English were not a constraint and a further 13 % felt that it was a slight constraint. These findings contradict those of Warren (2006) but the reason is that many use SMSs heavily and these can be in indigenous languages.

4.3.3 Socio-cultural Capital Outcomes

Contacting friends and relatives was the most significant advantage of using mobile phones reported. Nearly all (95 %) of the respondents indicated that mobile phones had significantly improved or greatly improved their relationships and contacts with friends and relatives. The mean score of 3.1 shows a highly positive response. The findings show that mobile phones had a positive impact on membership in groups or networks, a Social Capital aspect ($\mu = 3.0$). The findings support earlier studies (Goodman 2005; Kreutzer 2009) that found that the use of mobile phones contributes to improved social relationships. These findings suggest that mobile phones allow respondents to overcome communication challenges caused by long distance. Using mobile phones for communicating with family and friends outside the township area could reduce or eliminate travelling costs.

Almost two-thirds (65 %) of the respondents reported that mobile phone usage had either improved or greatly improved coordination of social activities. With the large numbers of social events, such as funerals and religious events in townships, it is not a surprise that mobile phones have contributed positively towards coordinating social events ($\mu = 3.1$). According to Sife et al. (2010), the use of mobile phones play a vital role in coordinating social events, subsequently reducing time and financial costs associated with arranging social functions. Over two-thirds (70.6 %) of the respondents indicated that mobile phones have contributed positively towards transportation and travelling issues. These results support the contention that mobile phones reduce the need to travel or simplify travel and transport arrangements, thereby saving time and money (Sife et al. 2010).

4.3.4 Socio-political Capital Outcomes

Mobile phone contribution towards communication with government departments was one of the lowest of the livelihood outcomes that were investigated ($\mu = 2.3$). It is clear that mobile phones have not yet played a significant role in improving communications with government departments. These findings could be as a result of lack of awareness among township dwellers about what the government could provide for them. These results support the argument that there is little evidence to show that communication with government has been improved by mobile phones (Reijswoud and Jager 2006; West 2006).

5 Conclusions

The overall findings of this study have shown that mobile phones contribute to township livelihood outcomes in respect of human and social capital but had little to no impact on financial and physical capital. The impact of mobile phones on human capital was shown in the positive contribution to receiving support in times of emergency and to the efficiency of daily activities. Education levels and lack of skills were not considered constraints on mobile phone use. Few feared using mobiles and did not find a lack of English a constraint. The impact on social capital was reflected in a positive contribution to relationships, contacts with friends and relatives and coordination of social events, and transportation and travelling arrangements.

Mobile phones were not considered to make any significant contribution to improving incomes of township dwellers, in sending and receiving money, in changing the status of the business, and in communication with government departments. The study identified a number of constraints facing township inhabitants in relation to mobile phone usage. Some of these identified were: high costs of mobile phones; low income; weak or unreliable networks and high costs of mobile phones. However, difficulties in getting airtime recharge vouchers were not one of the major constraints of using mobile phones in township areas.

This study's experiences suggest further research with larger sample sizes to help generalise the findings. In addition, research using a qualitative method could be conducted, where findings could be used to support and expand the findings of this study. Notwithstanding, this study has been able to show the relationship between mobile phones and livelihood outcomes.

Acknowledgments This work is based on the research supported in part by the National Research Foundation of South Africa (Grant Number 91022).

References

Adato M (2007) Methodological innovations in research on the dynamics of poverty: a longitudinal study on KwaZulu Natal, South Africa. World Dev 35(2):247–263

Adato M, Carter RM, May J (2006) Exploring poverty traps and social exclusion in South Africa using qualitative and quantitative data. J Dev Stud 42(2):226–247

Aker J, Mbiti I (2010) Mobile phones and economic development in Africa. J Econ Perspect 24 (3):207–732

Andrade AD, Urquhart C (2009) ICTs as a tool for cultural dominance. Electron J Inf Syst Dev Countries 27(2):1–12

Barbarin O, Khomo N (1997) Indicators of economic status and social capital in South African townships. Childhood 4(2):193–222

Baumann P (2002) Improving access to natural resources for the rural poor: a critical analysis of central concepts and emerging trends from a sustainable livelihoods perspective. FAO, LSP WP 1, Access to Natural Resources Sub-Programme

Bhavnani A, Chiu RW-W, Janakiram S, Silarszky P (2008) The role of mobile phones in sustainable rural poverty reduction. http://siteresources.worldbank.org/EXTINFORMATIONANDCOMMUNICATIONANDTECHNOLOGIES/Resources/The_Role_of_Mobile_Phones_in_Sustainable_Rural_Poverty_Reduction_June_2008.pdf. Accessed 28 June 2015

Buys P, Dasgupta S, Thomas T (2009) Determinants of a digital divide in sub-Saharan Africa: a spatial econometric analysis of cell phone coverage. World Dev 37(9):1494–1505

Chew H, Ilavarasan P, Levy M (2010) The economic impact of information and communication technologies (ICTs) on microenterprises in the context of development. Electron J Inform Syst Dev Countries 44(4):1–19

Diga K (2007) Mobile cell phones and poverty reduction: technology spending patterns and poverty level change among households in Uganda. Dissertation, University of KwaZulu Natal

Donner J (2007) The rules of beeping: exchanging messages via international "Missed calls" on mobile phones. J Comput Mediat Commun 13(1):1–22

Duncombe R (2006) Using the livelihoods framework to analyze ICT applications for poverty reduction through microenterprise. Massachusetts Inst Technol Inform Technol Int Dev 3 (3):81–100

Everatt D, Smith M (2008) Building sustainable livelihoods: an overview. National Department of Social Department, Pretoria, South Africa

Evoh J (2009) The role of social entrepreneurs in deploying ICTs for youth and community development in South Africa. J Community Inf 5(1):1–15

Fan S (2008) Public expenditures, growth, and poverty: lessons from developing countries. International Food Policy Research Institute, Washington, DC

Farrington J, Carney D, Ashley C, Turton C (1999) Sustainable livelihood in practice: early applications of concepts in rural areas. Overseas Dev Inst 42:43–73

Frempong G (2011) Developing information society in Ghana: how far? Electron J Inform Syst Dev Countries 47(8):1–20

Frye I, Magasela W (2005) Constructing and adopting an official poverty line for South Africa: some issues for consideration. National Labour and Economic Development Institute, Johannesburg

Goodman J (2005) Linking mobile phone ownership and use to social capital in rural South Africa and Tanzania. Vodafone policy paper series: Africa: the impact of mobile phones, pp 53–65

Gough N (2005) Africa: the impact of mobile phones—introduction. http://www.enlightenmenteconomics.com/assets/africamobile.pdf. Accessed 28 June 2015

Jack S, Suri T (2010) The economics of M-PESA. http://www.mit.edu/~tavneet/M-PESA.pdf. Accessed 12 Apr 2011

Kefela GT (2010) The impact of mobile phone and economic growth in developing countries. Afr J Bus Manag 5(2):269–275

Kenny C (2002) Information and communication technologies for direct poverty alleviation: cost and benefits. Dev Policy Rev 20(2):141–157

Kreutzer T (2009) Assesing cell phone usage in a South African township school. Int J Educ Dev Inf Commun Technol 5(5):43–57

Krueger AB, Maleckova J (2003) Education, poverty, political violence and terrorism: is there a causal connection? J Econ Perspect 17(4):119–144

Levine S (2014) How to study livelihoods: bringing a sustainable livelihoods framework to life, working paper 22. Secure Livelihoods Research Consortium, Overseas Development Institute, London

Maki M (2009) Addressing poverty in South Africa: an investigation of the Basic Income Grant. Dissertation, University of Pretoria

Masiteng K (2011) Quarterly labour force survey: quarter 4. Statistics South Africa, Pretoria

Mbiti I, Weil DN (2011) Mobile banking: the impact of M-Pesa in Kenya. http://www.econ.brown.edu/faculty/David_Weil/Mbiti%20Weil%20NBER%20working%20paper%2017129.pdf. Accessed 28 June 2015

McNamara K (2008) Enhancing the livelihoods of the rural poor through ICT: a knowledge map. Economic and Social Research Foundation (ESRF), Dar es Salaam

Ngugi B, Pelowski M (2010) M-pesa: a case study of the critical early adopter's role in the rapid adoption of mobile money banking in Kenya. Electron Inform Syst Dev Countries 43(3):1–16

Rao S (2009) Role of ICTS in India rural communities. J Community Inf 5(1):1–16

Rashid AT, Elder L (2009) Mobile phones and development: an analysis of IDRC-supported projects. Electron J Inform Syst Dev Countries 36(2):1–16

Reijswoud V, Jager A (2006) E-governance in the developing world in action: the case of DistrictNet in Uganda. J Community Inf 4(2):1–18

Samuel J, Shah N, Hadingham W (2005) Mobile communications in South Africa, Tanzania and Egypt: results from community and business surveys. Vodafone policy paper series. The Impact of Mobile Phones, Africa, pp 44–52

SARPN (2007) The measure of poverty in South Africa project: key issues. http://www.sarpn.org.za/documents/ d0002801/index.php. Accessed 9 Apr 2011

Shackleton S-J (2007) Rapid assessment of cell phones for development. http://www.unicef.org/southafrica/SAF_resources_cells4dev.pdf. Accessed 2 Apr 2011
Sife A, Kiondo E, Lyimo-Macha J (2010) Contribution of mobile phones to rural livelihoods and poverty reduction in Morogoro region, Tanzania. Electron J Inform Syst Dev Countries 42 (3):1–15
Statistics South Africa (2011) General household survey. Statistics South Africa, Pretoria
Statistics South Africa (2014) Poverty trends in South Africa: an examination of absolute poverty between 2006 and 2011. Statistics South Africa, Pretoria
Stillman L, Arnold M, Gibbs R, Shepherd C (2010) ICT, rural dilution and the new rurality: a case study of 'WheatCliffs'. J Community Inf 6(2):1–20
Studies in Poverty and Inequality Institute (2007) The measurement of poverty in South Africa project: key issues. SPII, Johannesburg
Tilak JB (2006) Post-elementary education, poverty and development in India. Int J Educ Dev 27 (4):435–445
Tshoose CI (2010) The impact of HIV/AIDS regarding informal social security: issues and perspectives from a South African context. PER 13(3):1727–1777
Warren K (2006) Globalization and health. Boston University School of Public Health, School of Public Health, Boston
West R (2006) Related dangers: the issue of development and security for marginalized groups in South Africa. J Community Inf 2(3):1–14

An Application to Support Sustainability Management in the Cuban Energy Sector

Frank Medel-González, Lourdes García-Ávila, and Jorge Marx Gómez

1 Introduction

Protecting the environment is one of the major challenges of businesses today. The majority of organizations are aware of their responsibility to the environment and society, however a limited number can turn their environmental and social strategic plans into action and can manage the timely generated sustainability information to facilitate the decision making process. The treatment of sustainability information and identifying a limited number of key indicators should help managers to make better decisions about corporate sustainability (CS) behavior. Information technologies (IT) can play an important role in sustainability management, specifically in the evaluation of sustainability performance (Medel-González et al. 2013).

The principal benefits of web applications are (Page and Rautenstrauch 2001):

- Increase the availability and the quality of data
- Decrease co-ordination efforts and time optimization
- Reduce time for data manual reports of different treatment
- Homogenize data structures
- Eliminate data redundancy

Web applications for information management can facilitate sustainability reporting and management and a number of examples can be observed (Freundlieb and Teuteberg 2012; Giesen et al. 2010; Johnson et al. 2014). In recent years there has been a range of techniques and frameworks developed that facilitate the development

F. Medel-González (✉) • L. García-Ávila
Universidad Central "Marta Abreu" de Las Villas, Santa Clara, Cuba
e-mail: frankmedel@uclv.edu.cu; lourdes@uclv.edu.cu

J. Marx Gómez
Carl von Ossietzky Universität, Oldenburg, Germany
e-mail: jorge.marx.gomez@uni-oldenburg.de

J. Marx Gómez, B. Scholtz (eds.), *Information Technology in Environmental Engineering*, Springer Proceedings in Business and Economics,
DOI 10.1007/978-3-319-25153-0_2

of dynamic web applications that can play a decisive role in the development of applications to support and manage the data generated by organizations. Despite these advances in Cuba, there are still gaps in relation to the assessment of sustainability performance as an internal management process in organizations. An additional gap is the failure in the information technologies support associated with the big data volumes to provide decision makers with a view of the progress and setbacks in business performance to assess the organization.

The main purpose of the paper is to present the design of a software application that supports the management of sustainability indicators and an overall index of sustainability performance evaluation, thereby allowing business performance evaluation and support for the process of sustainable decision-making.

2 Corporate Sustainability and Sustainability Management

Businesses have the responsibility in the process of transition to an improved sustainable development (SD). SD is a social concept and is being increasingly applied as a business concept under the concept of corporate sustainability (Steurer et al. 2005). The first definitions of corporate sustainability were a faithful translation of the concept given in "Our common future" report raised at corporate level. CS generally had the ability to allow organizations to meet their needs without compromising future needs of the organization and their stakeholders (Deloitte and Touche 1992; Hockerts 2001). Sustainability requires businesses to consider the environmental, social and economic aspects at the same time. If an organization is able to manage the risks and the opportunities holistically, it will lead to increased business success (Seidel 2013).

Managing corporate sustainability is a major challenge for companies to demonstrate their contribution to sustainable development in spite of the difficulties in measuring the performance of the corporate sustainability (Lee and Farzipoor Saen 2012). Sustainability management is the formulation, implementation and evaluation of both environmental and socioeconomic sustainability-related decisions and actions (Starik and Kanashiro 2013). The main objective of corporate sustainability management is balancing the organizational performance in the economic, social and environmental improvement opportunities identified simultaneously (Figge et al. 2002; Lee and Farzipoor Saen 2012; Macedo and Queiroz 2007; Schaltegger and Burritt 2005). Sustainability management includes the internal development of environmental and social measures as well as the external contribution to the sustainable development of society and the economy (Johnson 2015). Both concepts have an indissoluble relationship and have extreme significance to make the organizations more sustainable and reduce their negatives impacts and maximize the positive impacts.

Cuba is not immune to these international trends and in 2010 initiated the "National Program of consumption, sustainable production and efficient use of resources from 2010 to 2015". The main objective of the program was "… to promote changes in production, consumption and use of resources at the national level, in order to contribute to economic and social development on a sustainable basis . . ." (CITMA 2010). Other objectives of the program were to contribute to the sustainability and efficiency of Cuban business management. More recently, in 2011, the "Guidelines of the Economic and Social Policy of the Revolution" was released. Guideline 133 stated that to "Sustain and develop comprehensive research to protect, conserve and rehabilitate the environment and adapt the environmental policy to new economic and social projections" (PCC de Cuba 2011) was a priority. This shows the importance of the topic and the attention paid by the Cuban government. However, despite all efforts in Cuba, there are still large gaps in relation to sustainable performance evaluation as an internal management process for organizations, to facilitate the decision-making process based on key information associated with entrepreneurial behavior.

3 IT and Business Sustainability

Information technologies can play an important role in sustainability management, specifically in sustainability performance evaluation. Examples of the potential of IT include the collection of data on inputs and outputs of different processes, processing and storage of large volumes of data and the dissemination of information to different stakeholders (Page and Rautenstrauch 2001). In recent years software tools have been developed to help organizations in the process of sustainability reporting. Decision Support Systems are emerging as a suitable solution in the field of sustainability planning and control of complex systems (Filip 2008). Some of the most prominent specialized tools are: SAP Sustainability Performance Management (SuPM), Enablon SD-CSR, SoFi, credit360 and STORM. All of these tools presented offer similar functionality regarding their reporting capabilities (Rapp and Bremer 2013). Another important tool is OEPI, which is related to environmental performance indicators. A fundamental goal of OEPI is to bridge the gap between various sources and types of environmental information and users of different backgrounds by providing an integrated information source (Bracher 2013).

In Cuba, organizational information related to sustainability has become difficult to collect. The best results are in the field of environmental statistics in government official reports. The businesses need provide answers to key questions such as: what to measure? how to measure? and when to measure? These questions remain unanswered for many organizations and organizations indicate that they experience difficulties in obtaining the required information.

An additional problem is the information storage and availability which leads to the lack of IT support on sustainability performance evaluation. In recent years it has been an important concern in the Cuban business sector due to the fact that it hasn't been covered properly and it has inclusively found limitations in research and practical applications from the IT perspective.

4 Methodology

The research problem was identified as the lack of software tools in Cuba for sustainability assessment, to integrate consistent indicators related to the needs of company management and limitations in information technologies to support environmental and social data volumes, associated to environmental and sustainable performance. The System of Sustainability Performance Evaluation (SySPE) was designed to fill the gap in relation to the support of IT in the sustainability performance evaluation. The energy sector was selected in order to prototype the implementation of the application, SySPE in four power plants.

The research method to develop the application was a multi-methodological approach to Information System research called System Development (SD) (Burstein 2002; Nunamaker and Chen 1990). This method was used for the research through exploration and integration of available technologies to produce an artefact (Burstein 2002). According to Burstein (2002) and based on Nunamaker and Chen (1990) the systems development method consists of three steps: (1) concept building: investigating the functionality and requirements of the system and studying other disciplines for other ideas and approaches; (2) system building: the construction of the prototype system through the following steps: develop a system architecture, analyze and design the system and build the (prototype) system; and (3) system evaluation. SD can be useful to consider as part of the exploratory stage of an IS study.

5 Building the SySPE Prototype

The first stage of building the prototype was oriented to identify the functionality of the system. The objective of the application is to support sustainability management in distributed generation power stations in Cuba. The application utilizes a set of indicators distributed over the three pillars of sustainability, in a traditional Sustainability Balanced Scorecard perspective, to conform to a Corporate Index of Sustainability Performance (CISP). The CISP synthesizes the progress or setbacks in corporate sustainability performance in an index to verify in a simple and continuous way if the administration efforts, organizational management instruments and environmental training are transformed into a better sustainability performance. In order to make operative corporate sustainability measurement, a network structure should be defined by the organizations. The CISP design is based on the three sustainability pillars: (1) economic, (2) environmental and (3) social, all distributed over the four perspectives of SBSC. The indicators are grouped into the perspectives and could have relationship amongst them (Fig. 1).

The CISP (value range $0 \leq r_{ij} \leq 1$) formula is shown below (Formula 1) and it depends on three elements: (1) wp_i the relative weight of the perspective i, (2) wi_{ij} the relative weight of the indicator j in the perspective i and (3) r_{ij} Rate or

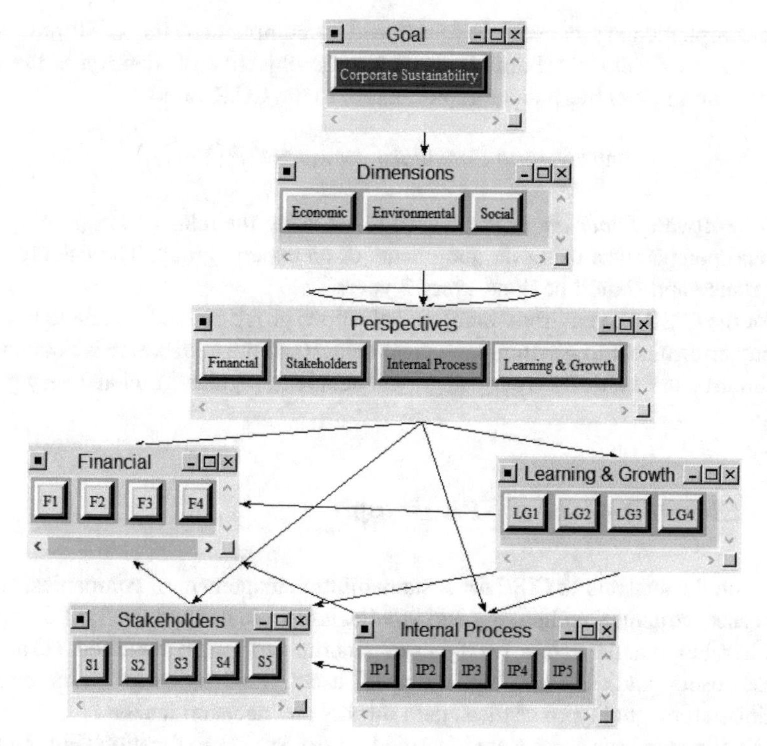

Fig. 1 CISP architecture with *Super-decisions* software (Medel-González et al. 2013)

normalized value of the indicator *j* of the perspective *i*. This formula measures the sum of contribution of each indicator in relation with their goal. CISP represents the accomplished level of the indicator goals in each perspective.

$$CISP = \sum_{i=1}^{4}\sum_{j=1}^{n} wp_i * wi_{ij} * r_{ij} \tag{1}$$

The r_{ij} (value range $0 \leq r_{ij} \leq 1$) is used to normalize different units among the indicators and made use of two values (1) x_{ij}: Numeric value of the indicator and (2) $Goal\{x_{ij}\}$ is the indicator defined goal (see Formula 2).

$$r_{ij} = \begin{cases} \dfrac{x_{ij}}{Goal\{x_{ij}\}} \text{ if } x_{ij} \text{ satisfies the condition "\textbf{more is better}"} \\ 1 \text{ if } x_{ij} \geq max\{x_{ij}\} \text{ "\textbf{more is better}"} \\ \dfrac{Goal\{x_{ij}\}}{x_{ij}} \text{ if } x_{ij} \text{ satisfies the condition "\textbf{less is better}"} \\ 1 \text{ if } x_{ij} \leq min\{x_{ij}\} \text{ "\textbf{less is better}"} \end{cases} \tag{2}$$

A complementary concept is introduced to complement the CISP and is the *Improvement Potentials* (Formula 3). It has the objective of identifying the most negative and influential indicators in relation to the CISP value.

$$Improvement\ Potentials_{ij} = wp_i{}^*wi_{ij}{}^*(1 - r_{ij}) \quad (3)$$

The software *Super decisions* is used to calculate the relative weight of indicators and perspectives using the judgments of an experts group. The calculus is of importance and should be done every 2 years.

For the CISP, the environmental organizations group defined a scale to translate the numeric value into a natural language: below 0.65 the CISP value is Poor and the maximum CISP value has four categories: deficient, regular, good and very good.

5.1 Elements of the SySPE Design

Based on the analysis of CISP for sustainability management of companies, actors, user cases, structure, technology and architecture were defined for SySPE application. In relation to the actors, two types of generics users were identified: (1) authenticated users and (2) not authenticated users. Authenticated users can be: administrators, quality specialists, data owners and decision makers.

The different user cases were identified for the SySPE application (Fig. 2). First, the administrator's role must be considered. This has the management and user management role associated with it. Secondly, the quality specialist's role relates to defining, deleting and modifying nomenclatures and indicators assigned to the

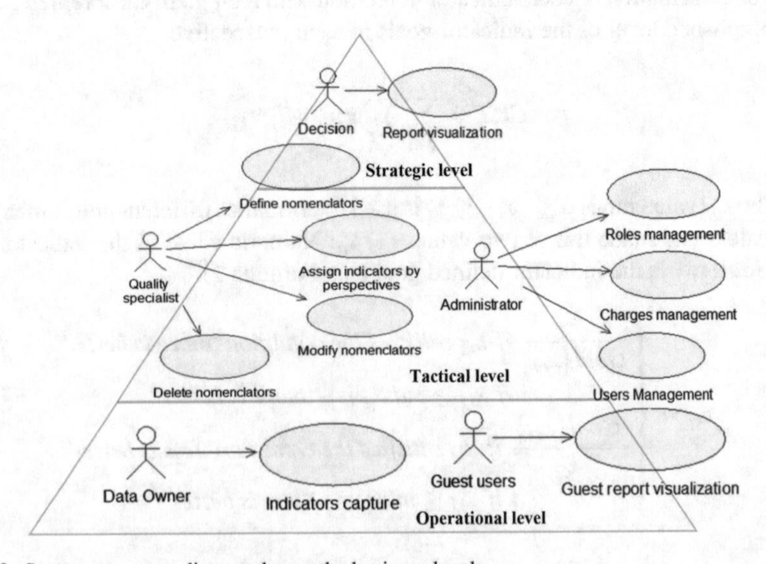

Fig. 2 System use cases distrusted over the business levels

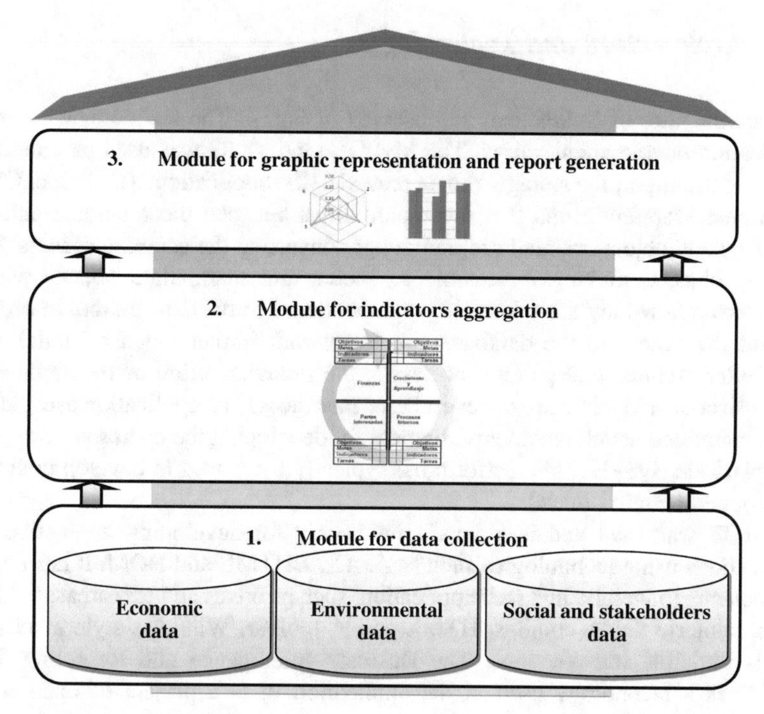

Fig. 3 Initial modules of SySPE application

perspectives. Thirdly, the data owner's role is connected with the capture of indicators and is responsible for recording the sustainability indicators. Finally, the decision maker is linked to the reports visualization generated by the system from the sustainability indicators and indices.

SySPE has three main modules (Fig. 3). The first one will be related to the collection and storage of indicators information defined by business managers, regulatory organizations and the experts group. The perspective definitions belong to this module too. The administrator should define the four perspectives of SBSC. Other functionalities are the update, modification or elimination of information. These actions will be restricted to a small group of users that could interact with the module system.

The second module will have the objective of setting the sustainability indicators defined over the SBSC perspectives. The second task is to assign weights to selected indicators and perspectives to calculate CISP. This module will be restricted to a small group of users. The third module allows users to visualize the behavior of CISP and sustainability indicators along a period and represent the behavior of indicators and indexes graphically. This module is expected to be used by the Business Intelligence and Reporting Tools (BIRT) for report generation.

5.2 *Architecture and Technologies*

The architecture of SySPE can be observed in Fig. 4. The technologies used for application development varies. The MySQL GUI v8.82 was used as a database manager to support the storage of data related to the application. The Propel Object Relational Mapping eliminates incompatibilities between the relational database language and object-oriented programming; converting the database schema XML in data objects, making it possible to access and manipulate objects without considering how they are related in correspondence to the data source. In order to control the access to the database, the framework implemented a Model View Controller architectural pattern, achieving the modularization of the application, to re-use code and make use of several user interfaces. The application uses Eclipse as an integrated development environment for developing the open source application platform SySPE. This platform has typically been used to develop integrated development environments.

Ext JS was used and is a JavaScript library for developing interactive web applications using technologies such as AJAX, DHTML and DOM. It has a set of components to include in a web application, such as boxes and text areas, fields for dates, numeric fields, combos, HTML editor, toolbar, Windows-style menus and panels divisible into sections. The Business Intelligence and Reporting Tools (BIRT) is a technology used in the application. It is a project of open source

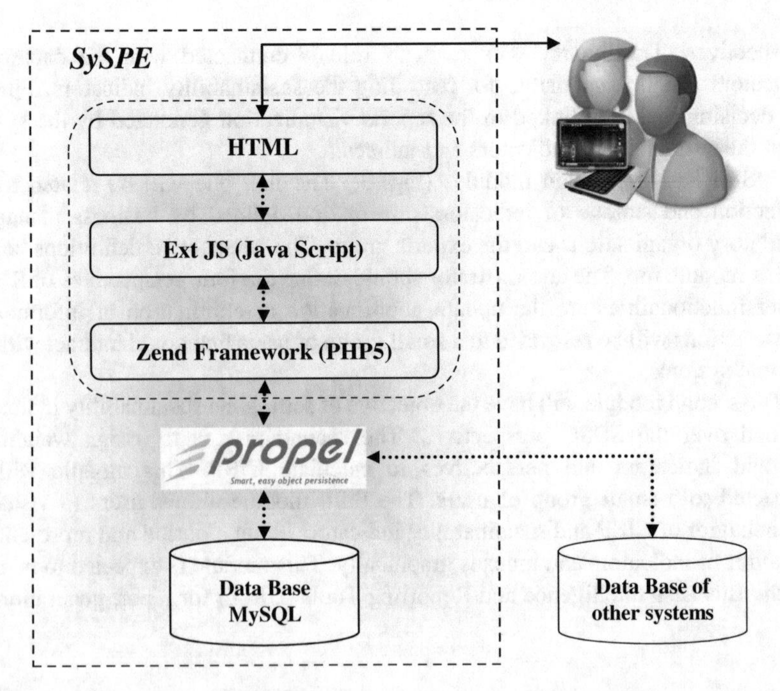

Fig. 4 Proposal architecture for SySPE

software that provides capabilities for reporting and business intelligence for web applications. BIRT also includes a graphics engine that is built into the report designer and can also be used separately to include graphics in an application.

Finally the XAMPP is used as server independent platform, free software which consists mainly of the MySQL database, Apache web server and interpreters for scripting languages: PHP and Perl. This program is licensed under the GNU web server acts as free, easy to use and capable of interpreting dynamic pages.

6 SySPE Application Interfaces

A number of aspects were taken into account for the interface design to facilitate the user's interaction (uses). The main menu contains all the possible applications functionalities and has the following elements:

- *Administration:* allows users to manage charges and roles
- *Nomenclatures*: in this menu can be defined the indicators, ranges (goals), indicators classifications, perspectives, activities, eco balances, impacts, strategic objectives, processes and risks are defined and included
- *Actions-risks:* propose actions to minimize environmental risks
- *Captures:* enable capturing records of predefined indicators
- *Perspective-indicator:* allows establishment of the relationship between the perspectives and the indicators and to set the relative importance of the indicators
- *Risk-activities:* assign the identified risks to the organization activities
- *Reports:* display reports defined in the application to assist decision makers

The system has sufficient error handling capabilities and avoids different mistakes in relation to the data capture activities. When the application detects an error, its displays a message with a red symbol and a concise text message is displayed that allows the user to identify the error.

Regarding the reports generation, different types of graphs can be displayed for the decision making process. The decision making users can consult all the indicators defined in the organization and their attributes like measure units, tendency (maximize or minimize indicators), goals and the last value of r_{ij} calculated. The perspectives contributions to the CISP can also be reported. It describes the organization's performance in the different areas such as: Financial, Stakeholders and Customers, Internal Process and Learning and Growth (Fig. 5a). Other reports of SySPE are the indicators related to the Stakeholders' perspective. These are shown in the dashboard graphics (Fig. 5b). Another important report is the improvement potentials. It offers the possibility to visualize the indicators who most affect the CISP (Fig. 5c). The application also provides the CISP calculation.

a)

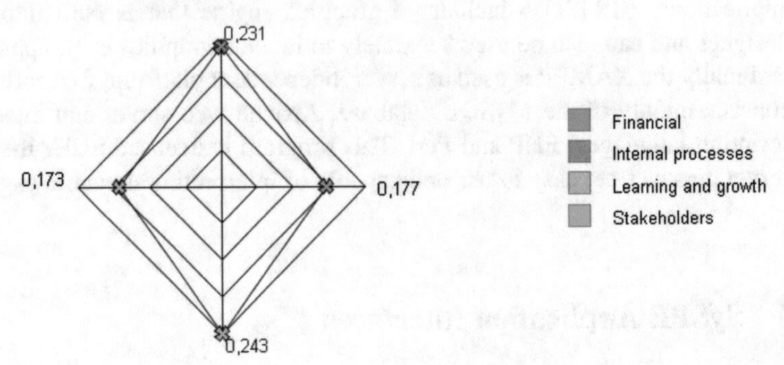

b)

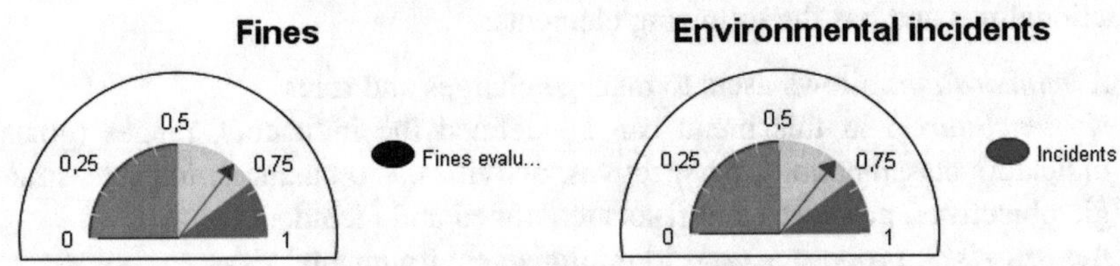

c)

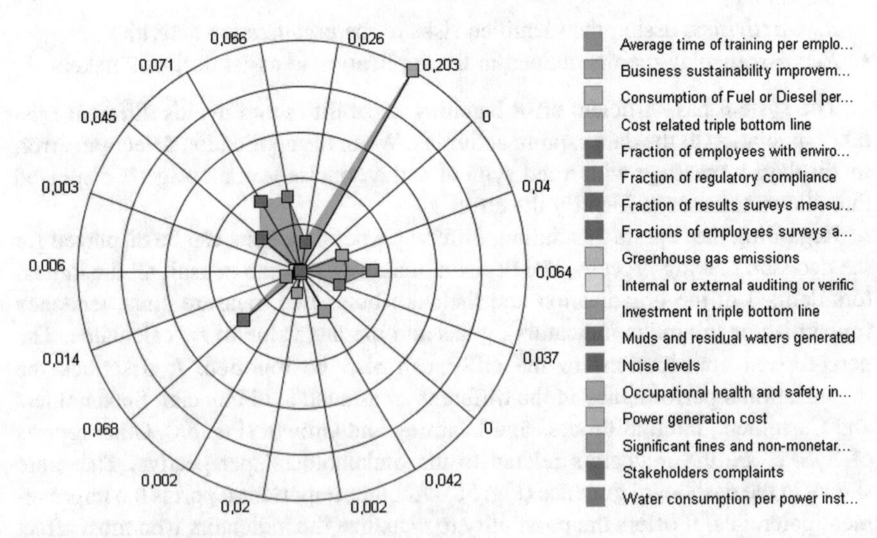

Fig. 5 Principal SySPE reports: (**a**) CISP contributions by perspective, (**b**) stakeholders indicators and (**c**) improvements potential by indicators

7 Limitations and Outlook

Currently the full implementation of the application is being completed. A primary objective is the steady generalization of data in the different business units, allowing information gathering associated with the organization's performance. The limitation of this paper is that it only covers two of the three steps of the research methodology and the evaluation of the systems will be probed by a case study of user behaviors and the implementation in the power plants.

In the future, the goal is to synchronize the SySPE application with the IT-for-Green project specifically in the third module related to Sustainability Reporting and Communication.

8 Conclusions

The integration of different technologies can help organizations to support the information related business sustainability performance and help the decision makers in sustainability management process as well. In order to support sustainability management in organizations, the design process allowed identifies the application roles, and user cases associated in different hierarchical levels of the organizations. A comprehensive and scalable architecture and applications modules were defined using accessible technologies.

SySPE provides a valuable tool to support the storage, retrieval and integration of different indicators, facilitating the calculation and graphical representation of CISP and improvement potentials. The application resolves one of the short comings of business sustainability performance evaluation in Cuban organizations.

Acknowledgments This work is funded by the Eureka SD project (agreement number 2013-2591), that is supported by the Erasmus Mundus programme of the European Union.

References

Bracher S (2013) OEPI platform. In: Dada A, Stanoevska K, Marx Gómez J (eds) Organizations' environmental performance indicators. Springer, Heidelberg, pp 103–116

Burstein F (2002) System development in information systems research. In: Williamson K (ed) Research methods for students, academics and professionals: information management and systems. Charles Sturt University, Wagga Wagga, pp 147–158

CITMA (2010) Programa nacional de consumo, producción sostenible y eficiencia en el uso de los recursos 2010–2015. Ministerio de Ciencia, La Habana

de Cuba PCC (2011) Lineamientos de la Política Económica y Social del Partido y la Revolución. Editora Política, La Habana

Deloitte & Touche (1992) Business strategy for sustainable development: leadership and accountability for the '90s. Business Strategy and the Environment. International Institute for Sustainable Development, Winnipeg

Figge F, Hahn T, Schaltegger S, Wagner M (2002) The sustainability balanced scorecard—linking sustainability management to business strategy. Bus Strateg Environ 11(5):269–284

Filip FG (2008) Decision support and control for large-scale complex systems. Annu Rev Control 32(1):61–70

Freundlieb M, Teuteberg F (2012) Evaluating the quality of web based sustainability reports: a multi-method framework. In: Proceedings of the 45th Hawaii international conference on system science, Maui, Hawaii, 4–7 Jan 2012

Giesen N, Jürgens P, Marx Gómez J, Omumi H, Rieken M, Stepanyan H, Süpke D (2010) ProPlaNET–Web 2.0 based sustainable project planning. In: Greve K, Cremers AB (eds) EnviroInfo 2010: integration of environmental information in Europe. 24th international conference on informatics for environmental protection in cooperation with InterGeo 2010, Cologne-Bonn, Oct 2010. Shaker, Aachen, pp 436–445

Hockerts K (2001) Corporate sustainability management—towards controlling corporate ecological and social sustainability. In: Proceedings of 9th international conference of greening of industry network, Bangkok, 21–25 Jan 2001

Johnson MP (2015) Sustainability management and small and medium-sized enterprises: managers' awareness and implementation of innovative tools. Corp Soc Responsib Environ Manag 22(5):271–285. doi:10.1002/csr.1343

Johnson M, Viere T, Schaltegger S, Halberstadt J (2014) Application of software and web-based tools for sustainability management in small and medium-sized enterprises. In: Proceedings of the 28th enviroinfo 2014 conference, Oldenburg, Germany, 10–12 Sept 2014

Lee K-H, Farzipoor Saen R (2012) Measuring corporate sustainability management: a data envelopment analysis approach. Int J Prod Econ 140(1):219–226

Macedo AVP, Queiroz ME (2007) Gerenciando e Otimizando a Sustentabilidade Empresarial através da Ferramenta Balanced Scorecard: Em Busca da Mensuração. In: Congresso Virtual Brasileiro de Administração—CONVIBRA, 7–15 Dec 2007

Medel-González F, García-Ávila L, Acosta-Beltrán A, Hernández C (2013) Measuring and evaluating business sustainability: development and application of corporate index of sustainability performance. In: Erechtchoukova MG, Khaiter PA, Golinska P (eds) Sustainability appraisal: quantitative methods and mathematical Techniques for environmental performance evaluation. Ecoproduction (Environmental issues in logistics and manufacturing). Springer, Heidelberg, pp 33–61

Nunamaker JF Jr, Chen M (1990) Systems development in information systems research. In: Proceedings of the 23rd annual Hawaii international conference on system sciences, Kailua-Kona, Hawaii, 2–5 Jan 1990

Page B, Rautenstrauch C (2001) Environmental informatics—methods, tools and applications in environmental information processing. In: Patig S, Rautenstrauch C (eds) Environmental information systems in industry and public administration. Idea Group Publishing, Hershey, pp 2–11

Rapp B, Bremer J (2013) IT solutions for EPI management. In: Dada A, Stanoevska K, Marx Gómez J (eds) Organizations' environmental performance indicators. Springer, Heidelberg, pp 19–31

Schaltegger S, Burritt R (2005) Corporate sustainability. In: Folmer H, Tietenberg T (eds) The international yearbook of environmental and resource economics 2005/2006: a survey of current issues. Edward Elgar Publishing, Cheltenham, pp 185–222

Seidel S (2013) Interview with daniel schmid on "sustainability and the role of IT". Bus Inf Syst Eng 5(5):327–329

Starik M, Kanashiro P (2013) Toward a theory of sustainability management: uncovering and integrating the nearly obvious. Organ Environ 26(1):7–30

Steurer R, Langer ME, Konrad A, Martinuzzi A (2005) Corporations, stakeholders and sustainable development I: a theoretical exploration of business–society relations. J Bus Ethics 61 (3):263–281

A Framework for Environmental Management Information Systems in Higher Education

Brenda Scholtz, André P. Calitz, and Blessing Jonamu

1 Introduction

Higher Education Institutions (HEIs) have been called on not to be bystanders as the world faces increasing environmental issues and to take action and be pioneers in driving environmental sustainability into the community (Savely et al. 2007). There has been increased interest in the topics of environmental sustainability and management of environmental information at HEIs. This is particularly evident in Europe, the United States (U.S.), Asia, Australia, Canada and South America where several studies (Disterheft et al. 2012; Fonseca et al. 2011; Jones et al. 2011; Velazquez et al. 2006) have reported efforts at reducing environmental impact in HEIs. This could be as a result of the pressure which is being placed on these institutions by Environmental Protection Agencies, stakeholders, governments and non-government organisations to act responsibly towards the environment (Savely et al. 2007).

This internal and external pressure on HEIs has prompted them to follow the industry trend of developing environmental programs and systems (Bero et al. 2012). The United States of America's Environmental Protection Agency (EPA) highlighted that the environment has become everyone's business and everyone now has a right to access high quality environmental information (EPA 2009). A key factor for achieving environmental sustainability is the management of this environmental information (Bero et al. 2012; Speshock 2010). HEIs have a unique set of challenges with regards to environmental information management and these are different to industrial contexts (Bero et al. 2012; Kamal and Asmuss 2013). The benefits of having environmental information readily available include well informed and improved decision making which facilitate the reduction in the

B. Scholtz (✉) • A.P. Calitz • B. Jonamu
Department of Computing Science, Nelson Mandela Metropolitan University (NMMU), Port Elizabeth, South Africa
e-mail: brenda.scholtz@nmmu.ac.za; andre.calitz@nmmu.ac.za

J. Marx Gómez, B. Scholtz (eds.), *Information Technology in Environmental Engineering*, Springer Proceedings in Business and Economics,
DOI 10.1007/978-3-319-25153-0_3

environmental footprint of the institution, in terms of a reduction in water usage, an improvement in energy efficiency and a reduction in pollution (Jones et al. 2011). It can also raise the environmental awareness of the institution's community which includes students, staff members and external stakeholders.

According to Velazquez et al. (2006), a HEI is considered to be a sustainable institution if the institution addresses, involves and promotes the minimisation of negative environmental, economic, social and health impact of daily activities involved in the functioning of the institution. Sustainable universities have influence to the extent that they help regional or global societies to transition to a sustainable livelihood. Whilst the importance of environmental sustainability and the relevant role of environmental information management and reporting in higher education has been emphasised in HEIs, until recently Africa has been lagging behind (Bosire 2014) and research studies in this field are limited. In South Africa there is escalating pressure on HEIs to report on sustainability since the Department of Higher Education and Training recently published a government notice dictating that it is mandatory that public HEIs report on environmental impact effective from the year 2015 (Department of Higher Education and Training 2014).

Most aspects of environmental sustainability are tightly dependent on the availability and accessibility of correct and current environmental information (El-Gayar and Fritz 2006). Organisation stakeholders need access to environmental information to evaluate and assess the environmental dimension of organisational decisions, both at a managerial level and at a strategic level. Hence, environmental sustainability requires correct Environmental Information Management (EIM). However a lack of coordinated effort and poor decision making with regards to achieving environmental sustainability in HEIs has been reported in some studies (Bosire 2014; Jones et al. 2011). This poor decision making can be attributed to the lack of efficient processes, structures and information systems which support centralised environmental data sources and eliminate information silos in different departments, faculties and campuses (Bero et al. 2012; Jones et al. 2011; Velazquez et al. 2005).

The most predominant initiative to enhance environmental sustainability in HEIs is the implementation of Environmental Management Systems (EMS) at these institutions (Disterheft et al. 2012; Jones et al. 2011). The ISO 14001 standard, specified by the International Standards Organisation (ISO), defines an EMS as a part of a management system that consists of planning activities, processes, procedures and resources for developing and maintaining of environmental policies within an organisation (ISO 2004). An EMS is not a computer system but rather a set of management tools and principles designed to aid an organisation to incorporate environmental concerns in their daily business activities (Speshock 2010).

An Environmental Management Information System (EMIS) is defined by El-Gayar and Fritz (2006) as an 'organizational-technical system for systematically obtaining, processing, and making relevant environmental information available in companies'. However, several studies (Bero et al. 2012; Savely et al. 2007) show that diverse and manual systems are still evident in HEIs. Other problems include the use of paper based records for resource usage, data quality issues and lack of data and process integration (Bero et al. 2012; Scholtz et al. 2014).

Whilst several studies (Athanasiadis 2006; Bero et al. 2012; Giesen et al. 2009; Solsbach et al. 2010) have proposed an EMIS, some of which are specifically designed for HEIs, each of these have a different focus or purpose. In addition there is limited research which provides formal guidance and best practice regarding the design of these EMIS. Studies of environmental sustainability and the role of information management in African HEIs are limited. The aim of this paper is therefore to propose a framework for the design of an EMIS for HEIs and to incorporate associated guidelines. The guidelines are based on best practice from literature and are classified according to the common components and responsibilities of an EMIS identified by several authors. This paper also highlights the increasing importance of environmental information and EMIS to environmental sustainability efforts at HEIs.

The next section provides a concise description of the research objectives and the methodology that was employed in this study. This is followed in Sect. 3 by a rigorous literature review that highlights that the design of an effective EMIS is critical to any attempt at reducing any negative environmental impact. The analysis of literature identified several common components of EMIS and resulted in the proposed theoretical framework for EMIS at HEIs. The last section provides several conclusions and recommendations for future research.

2 Research Objectives and Methodology

It has been reported that there is limited formal guidance or frameworks for HEIs that wish to adopt environmental sustainability practices, particularly for the management of environmental information (Bero et al. 2012). The main research objective of this paper is therefore: "*To propose and implement a framework that supports the design of an efficient EMIS at an HEI.*" In order to derive this model, two secondary objectives must be achieved:

- Propose an architectural framework for EMIS in higher education
- Propose guidelines for the components and responsibilities of an EMIS

In order to address the objectives and produce the artefacts (the framework and guidelines), this research used the Design Science Research (DSR) methodology which is used frequently in IS' projects producing artefacts (Hevner 2007). A careful rigorous investigation into literature, previous research and extant systems was done in order to gain an understanding of the research domain. This allowed for the identification of some of the key components of an EMIS and their responsibilities as well as guidelines for the design and implementation of an EMIS. The theoretical guidelines can be used to guide the design of an EMIS.

3 Components of Environmental Management Information Systems

The adoption of EMISs as tools to support environmental efforts has been found to be most prominent in industries that have a significant impact on the environment. Such industries include pharmaceuticals, oil, hazardous chemicals, automotive, utilities, primary metals and semiconductors industries. The challenges faced by HEIs are slightly different to those of industrial contexts and are (Alshuwaikhat and Abubakar 2008; Bero et al. 2012; EPA 2007):

- An extremely diverse community of faculty, students, staff, and support personnel, all with differing priorities, modes of engagement, and supervisory models
- A broad range of institutional activities and facilities including offices, laboratories, machines, classrooms, dining halls, and dormitories
- Broad distribution across a range of buildings and facilities of differing design and age, potentially dispersed over a large area
- Relatively limited financial and personnel resources for developing, implementing, and sustaining an effective EMS and EMIS

Any sustainability efforts must include a planning phase where the institution's stakeholder's such as top management and the IT department establish policies and objectives of their EMS (UNECE 2014). At this stage the environmental indicators (for example, energy and water) must be identified and prioritised. The Global Reporting Initiative (GRI) is a popular standard for establishing environmental indicators in industry (GRI 2013). Non-institutional organisations usually have a main focus which determines their environmental indicators, for example pharmaceutical companies would have their environmental indicators focused on hazardous waste. By contrast, HEIs have a broad set of institutional activities and facilities including offices, laboratories, operating machinery, classrooms, dining halls, dormitories, and maintenance, hence environmental indicators associated with HEIs are generally more diverse. The ISO14000 set of standards produced by the International Standards Organisation (ISO 2004) is one which is frequently used in universities in the U.S. and Europe (Alshuwaikhat and Abubakar 2008; Jones et al. 2011). The Sustainability Tracking, Assessment and Rating System (STARS) is a voluntary, self-reporting framework for recognising and gauging sustainability performances specifically for the higher education environment (AASHE 2012). The four main focus areas of an HEI identified by STARS are (1) education and research; (2) operations; (3) planning, administration and engagement and (4) innovation. Thus, the complex structure of HEIs requires that EMISs be tailored for HEIs (Kamal and Asmuss 2013).

Once the planning stage of an EMS has been completed, the environmental information architecture must be determined (Speshock 2010). Muntean et al. (2010) propose an architectural framework for a Performance Management System (PMS) for HEIs. Such a system should provide the instruments to support the governance processes, showing the data and analysis necessary for strategic planning and control. The four main components or layers of the PMS framework are: (1) the data layer; (2) the reporting layer; (3) the analytical layer and (4) the monitoring layer. The PMS framework uses data warehousing technology to Extract, Transform and Load (ETL) the data into the lowest layer which is the data layer containing the university data warehouse (Muntean et al. 2010). The ETL processes will then allow for data aggregation, normalisation and integration. Data is extracted from various sources and is stored in the database of the data warehouse which is in the data layer. The reporting layer allows users to access and query data. In addition, the reporting layer allows for ad-hoc querying and standard report generating from the university portal and is valuable for managerial decision making. The analytical layer is a useful tool for management in decision making and strategising. This layer allows for advanced functionality such as data mining, Online Analytical Processing (OLAP), forecasting, multidimensional/OLAP analysis, data mining, text mining, forecasting, decision support and predictive modeling. The monitoring layer is for performance monitoring of data from the various pools of data where meaningful information is displayed by means of performance dashboards and scorecards. A performance dashboard is an application that allows stakeholders to measure, monitor and manage organisation performance more effectively. The university portal forms part of the presentation layer which is the hub of all the university IT applications and services needed by students, administrators, faculty and staff (Bero et al. 2012; Muntean et al. 2010).

A further argument for extending the PMS framework to the domain of environmental sustainability is that its architecture has similar components to those of an EMIS identified by several studies (Bero et al. 2012; El-Gayar and Fritz 2006; Giesen et al. 2009; Gunther et al. 2004). The study of Gunther et al. (2004) also uses data warehousing technology and has been successfully used for analysing and querying environmental information.

An EMIS can be described as the system that maintains and enhances the environmental knowledge base of a company in order to meet the information needs of its environmental professionals (Al-Ta'ee et al. 2013). An EMIS has also been defined as the backbone or a precondition to environmental management efforts which supports the organisation's EMS and meets the reporting needs of stakeholders (El-Gayar and Fritz 2006). Accordingly, the definitions of EMIS are very broad and can incorporate a broad selection of systems and components from stand-alone end-user systems (for example, spreadsheets) to more complex, intelligent systems to enterprise wide integrated IS and can include performance management (Bero et al. 2012).

Typically an EMIS is implemented in a large organisation where there are several data sources which are characteristically located in different physical locations and diverse implementations (El-Gayar and Fritz 2006). These organisations already have other IS in place to automate aspects of the organisation, for example legacy systems and/or Enterprise Resource Planning (ERP) systems. A comparison of three EMISs revealed several common features and differences. The Adaptive Intelligent Service Layer for Environmental information management (AISLE) is one such service-oriented EMIS that mediates between environmental data providers and actual end user applications that require pre-processed environmental information (Athanasiadis 2006). The Sustainable Online Reporting Model (STORM) is a web-based EMIS that is used mainly for sustainability reporting (Solsbach et al. 2010). Sustainability reporting efforts are tightly associated with EMISs just as much as environmental management efforts are. Sustainability reporting requires that environmental information is retrieved from the various information sources and EMISs serve this particular purpose in sustainability reporting efforts. STORM seeks to address such issues and can retrieve data for reporting from legacy information systems and other databases or sensor networks.

All three systems investigated (STORM, AISLE and DEMS) have data collection capabilities which include pulling data from various sources. However, the level of capability within the data collection component varies. The AISLE system focuses on providing high quality data hence it performs extensive data pre-processing as data is collected. The DEMS has the capability of distributed manual entry of data. All three have a central database in which they store data. However, STORM and DEMS do not allow for access to this raw data but AISLE can provide access to the data to third party end-user applications which then process the data to provide valuable information. The main focus of STORM is for external, stakeholder sustainability reporting. It is thus evident that the objectives for EMIS can vary.

An analysis of several studies of environmental information efforts and EMIS design (Bero et al. 2012; El-Gayar and Fritz 2006; Giesen et al. 2009; Muntean et al. 2010; Su et al. 2013) reveal several commonly identified components (Table 1). These components can be classified into five commonly identified categories of components of an EMIS, namely:

- Data collection
- Centralised data storage and access
- Data processing and analysis
- Reporting (ad-hoc querying)
- Monitoring

The first component category of an EMIS is the data collection component which includes the integration of legacy and heterogeneous systems (Bero et al. 2012; Giesen et al. 2009). Data collection mechanisms vary depending on the legacy systems or the lack of legacy systems but can include those used for data acquisition and data pre-processing such as data cleaning, validation, integration and normalisation. Data from automated systems with environmental sensors or smart meters can be retrieved automatically and directly from the systems and streamed to the appropriate tables in the data layer of the EMIS (Bero et al. 2012; Su et al. 2013). However, many environmental metrics cannot be collected by automated systems and must be entered onto paper based records and then captured manually, often leading to duplication in effort and errors (Bero et al. 2012). A goal of an efficient EMIS is to streamline distributed manual data collection "at the source" using specialised web-based interfaces or with a handheld tablet or device. Spatial data types such as transportation and parking must also be accommodated.

Table 1 Components of EMIS

Component category	Component	Authors
Data collection	Automated data collection (sensors, sensor networks, smart meters)	Bero et al. (2012), El-Gayar and Fritz (2006), Su et al. (2013)
	Distributed data entry at location	Bero et al. (2012)
	Legacy system integration, connectivity to enterprise or ERP systems	Bero et al. (2012), Giesen et al. (2009)
	Data cleaning, validation and verification Integration and normalisation Extract, Transform and Load (ETL)	Athanasiadis (2006), Solsbach et al. (2010), Speshock (2010), Su et al. (2013)
Centralised data storage and access	Data layer (storage layer; data warehouse layer)	Athanasiadis (2006), Bero et al. (2012)
Reporting	Reporting layer (querying and reporting)	El-Gayar and Fritz (2006), Giesen et al. (2009), Solsbach et al. (2010), Su et al. (2013)
Data processing and analysis	Analytical layer (Analysis/analytical tools, OLAP, data mining and forecasting)	Al-Ta'ee et al. (2013)
	Aggregation, simulation, modelling of data and decision support	Bero et al. (2012), El-Gayar and Fritz (2006), Giesen et al. (2009), Speshock (2010)
Monitoring	Monitoring layer (dashboards and scorecards)	Bero et al. (2012)
Presentation	Presentation layer (public access)	Su et al. (2013)

One other key role that EMISs play in sustainability reporting is the verification of the environmental information to be published (Solsbach et al. 2010). The data also needs to be validated at the source, integrated and normalised where necessary. ETL processes such as cleaning and integrating also need to take place to prepare the data for the data warehouse.

Centralised data storage and access is another commonly identified component in an EMIS (Athanasiadis 2006; Bero et al. 2012) and in the PMS architecture for HEIs proposed by Muntean et al. (2010). After the retrieval of the data from the various data sources, data can be stored and processed to produce meaningful information (Athanasiadis 2006). Some EMIS are also developed to cater for document management since in environmental efforts documents such as environmental policies need to be stored in a safe and secure environment. The reporting component is also an important component and essential to an EMIS (El-Gayar and Fritz 2006; Giesen et al. 2009; Solsbach et al. 2010; Su et al. 2013). Reporting of sustainability information can foster public participation, social responsibility and promotion of sustainability in teaching and research (Alshuwaikhat and Abubakar 2008).

Another commonly identified component category and responsibility of an EMIS is data processing and analysis. Organisations need to process environmental data into useful information which can be used to draw meaningful conclusions (Speshock 2010). Advanced EMIS offer the capability to analyse environmental information, simulate and provide decision support (El-Gayar and Fritz 2006). These capabilities are useful and make an EMIS valuable to the top management of any institution. Data processing further involves complex algorithms that provide aggregation, ad-hoc querying and modelling of environmental data and processes (Bero et al. 2012). This is referred to as the analytical layer in the Muntean framework which allows management to make decisions and strategise and allows for advanced functionality such as data mining, OLAP and forecasting. The monitoring layer is for performance monitoring. Tools that are available in this layer include dashboards and scorecards (Bero et al. 2012; Muntean et al. 2010). Data is then analysed and produced for information distribution to stakeholders through a presentation layer (Su et al. 2013) where there should be support for public awareness and outreach by allowing access to simplified aggregated data summaries of data for access by the public and HEI stakeholders (for example, student, staff, board members, management and government bodies).

Athanasiadis (2006) divides the components and features of a EMIS into three clusters of services, namely: the contribution services cluster, the management services cluster and the distribution services cluster. Based on the definition of these clusters and an analysis of their responsibilities, the four common categories components of an EMIS can be classified into these clusters. Data collection and centralised data storage can be classified in to the contribution services cluster while data processing, monitoring and reporting fall under the managerial services cluster. The distribution services cluster is responsible for access to data by stakeholders and presentation of data. In the contribution services cluster there needs to be support for allocating resource usage to buildings and campus facilities such as sports grounds and departments (Bero et al. 2012). Apportionment modifiers are used to algorithmically apportion meter utility usage to arbitrary buildings and spaces within a metered loop (Bero et al. 2012). This data is stored in the space/location entity in the data source layer. The resulting framework can be used to design an EMIS for HEIs and shows all the components (Fig. 1). The related guidelines for the components of an EMIS are summarised in Appendix.

4 Conclusions and Recommendations

This paper proposed a framework which can assist HEIs to improve their environmental information management efforts and provides guidelines for the components of an EMIS which can therefore assist with designing an EMIS. An improvement in EIM can facilitate on improvement in decision making, environmental awareness and community involvement. This study forms part of a larger research study which aims to design and develop a university-wide environmental information data warehouse. The development of the data warehouse is focused on making environmental data accessible for querying and to end user applications.

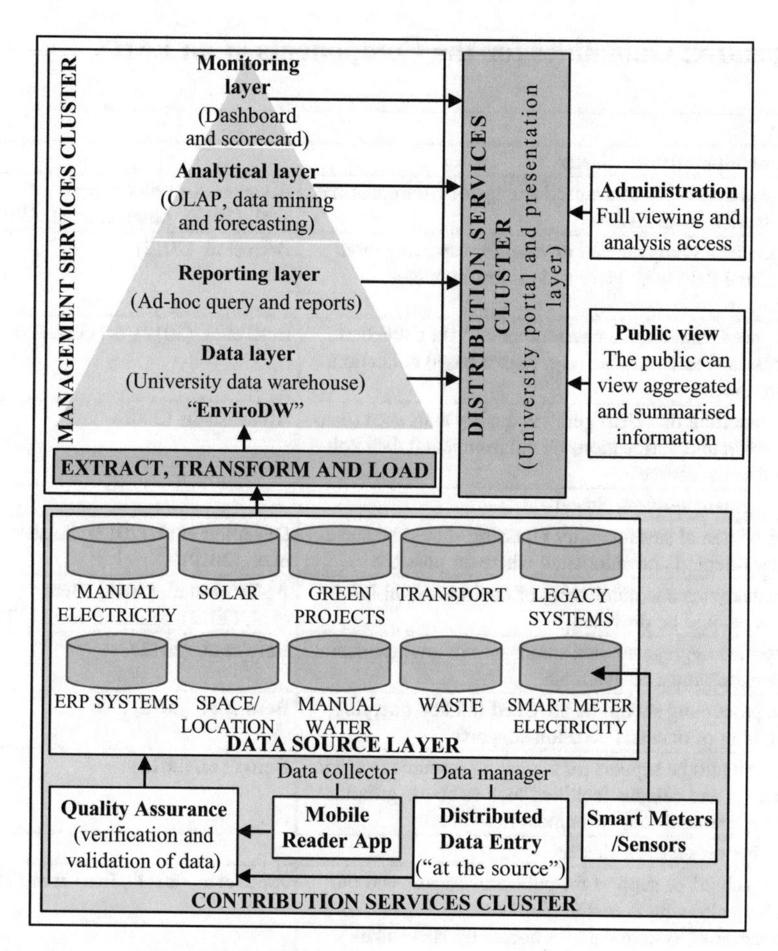

Fig. 1 A framework for EMIS design in HEIs

This will allow for the development of innovative end user applications to serve as the portal for which stakeholders can access the environmental information and view it in a meaningful way. This research is to serve as proof-of-concept in a hope to drive senior management to support and drive towards creating an environmentally sustainable institution and an inclusive community.

The development of the fully analytical tools to support senior management in decision management is important. Future research can be done to evaluate this theoretical framework in other higher education institutions. Research is required regarding the adoption of EMIS and criteria for evaluating these systems. Improving the effectiveness of EMIS and improving the changes of successful implementation will ultimately improve the success of environmental efforts. Generally HEIs are publically funded therefore transparency regarding environmental impact is of utmost importance.

Appendix: Guidelines for the Components of an EMIS

Guideline	Authors
Contribution services cluster	
There is a need for centralised storage of environmental information in an HEI	Athanasiadis (2006), Bero et al. (2012), Muntean et al. (2010)
HEIs need a computerised process for capturing environmental data (e.g. water and electricity meter readings)	Bero et al. (2012)
HEIs need to automate environmental data collection processes where possible (e.g. sensor based collection; smart meters)	Bero et al. (2012), Su et al. (2013)
It is important that HEIs perform quality assurance (data validation and verification) on environmental data collected at the source	Athanasiadis (2006)
Management services cluster	
Environmental sustainability reporting should be stakeholder oriented and automated wherever possible	Disterheft et al. (2012), Solsbach et al. (2010)
Data analytics and monitoring of environmental information should be provided	Al-Ta'ee et al. (2013), Bero et al. (2012), Muntean et al. (2010)
Simplified aggregated data summaries of environmental information must be available	Bero et al. (2012)
Data processing should be provided (ad-hoc querying, modelling of data and decision support)	Bero et al. (2012)
There should be support for allocating resource usage to buildings and campus facilities such as sports grounds and departments (apportionment modifiers)	Bero et al. (2012)
Distribution services cluster	
There should be support for public awareness and outreach by allowing access to simplified aggregated data summaries of system data for access by HEI stakeholders (student, staff, board members, management, government bodies, etc.)	Jones et al. (2011), Bero et al. (2012)
Third party applications should be granted access to environmental information where possible	Athanasiadis (2006), Bero et al. (2012)

References

AASHE (2012) STARS: version 1.2 technical manual. Association for the advancement of sustainability in higher education. AASHE, Denver

Alshuwaikhat HM, Abubakar I (2008) An integrated approach to achieving campus sustainability: assessment of the current campus environmental management practices. J Clean Prod 16 (16):1777–1785

Al-Ta'ee M, El-Omari NK, Ghwanmeh S (2013) Innovative study utilizing EMIS in supporting participatory urban decision making process in Jordan. Int J Appl Inf Syst 5(2):1–13

Athanasiadis IN (2006) An intelligent service layer upgrades environmental information management. IT Prof 8(3):34–39

Bero BN, Doerry E, Middleton R, Meinhardt C (2012) Challenges in the development of environmental management systems on the modern university campus. Int J Sustain High Educ 13(2):133–149

Bosire S (2014) A sustainability reporting framework for a higher education Institution. Dissertation, Nelson Mandela Metropolitan University

Department of Higher Education and Training (2014) Regulations for reporting by public higher education institutions. Government Gazette 37726

Disterheft A, da Silva Caeiro SS, Ramos MR, de Miranda Azeiteiro UM (2012) Environmental Management Systems (EMS) implementation processes and practices in European higher education institutions—top-down versus participatory approaches. J Clean Prod 31:80–90

El-Gayar O, Fritz BD (2006) Environmental Management Information Systems (EMIS) for sustainable development: a conceptual overview. Commun Assoc Inf Syst 17:756–784

EPA (2007) Environmental Management Systems (EMSs) for colleges and universities. http://www.epa.gov/region1/assistance/univ/emsguide.html. Accessed 05 July 2014

EPA (2009) Information access strategy. http://www.epa.gov/nationaldialogue/FinalAccessStrategy.pdf. Accessed 29 June 2014

Fonseca A, Macdonald A, Dandy E, Valenti P (2011) The state of sustainability reporting at Canadian universities. Int J Sustain High Educ 12(1):22–40

Giesen N, Farzad TH, Marx Gómez J (2009) A component based approach for overall Environmental Management Information Systems (EMIS) integration and implementation. Shaker Verlag, Berlin

GRI (2013) Indicator protocols set environment (EN). https://www.globalreporting.org/resourcelibrary/G3.1-Environment-Indicator-Protocols.pdf. Accessed 10 May 2014

Gunther S, Marx Gómez J, Rautenstrauch C (2004) Modeling of a data warehouse system for environmental information. In: Proceedings of the world automation congress, Seville, Spain, 28 June–1 July 2004

Hevner AR (2007) A three cycle view of design science research a three cycle view of design science research. Scand J Inf Syst 19(2):1–5

ISO (2004) ISO 14000—environmental management. http://www.iso.org/iso/home/standards/management-standards/iso14000.htm. Accessed 04 June 2014

Jones N, Panoriou E, Thiveou K, Roumetiotis S, Allan S, Clark JRA, Evangelinos KI (2011) Investigating benefits from the implementation of Environmental Management Systems in a Greek university. Clean Techn Environ Policy 14(4):669–676

Kamal ASM, Asmuss M (2013) Benchmarking tools for assessing and tracking sustainability in higher educational institutions: identifying an effective tool for the University of Saskatchewan. Int J Sustain High Educ 14(4):449–465

Muntean M, Sabau G, Bologa A, Surcel T, Florea A (2010) Performance dashboards for universities. In: Proceedings of the 2nd international conference on manufacturing engineering, quality and production systems, Constantza, Romania, 3–5 Sept 2010

Savely SM, Carson AI, Delclos GL (2007) An environmental management system implementation model for U.S. colleges and universities. J Clean Prod 15(7):660–670

Scholtz B, Calitz A, Jonamu B (2014) A gap model for environmental information management in an African Higher Education Institution. In: Proceedings of the 8th IDIA international development informatics association conference, Port Elizabeth, South Africa, 3–4 Nov 2014

Solsbach A, Supke D, Wagner vom Berg B, Marx Gómez J (2010) Sustainability online reporting model—a web based sustainability reporting software. In: Golinska P, Fertsch M, Marx Gómez J (eds) Information technologies in environmental engineering, new trends and challenges, vol 3, Environmental science and engineering. Springer, Heidelberg, pp 165–177

Speshock CH (2010) Empowering green initiatives with IT: a strategy and implementation guide. Wiley, New Jersey

Su X, Shao G, Vause J, Tang L (2013) An integrated system for urban environmental monitoring and management based on the environmental internet of things. Int J Sust Dev World 20 (3):205–209

UNECE (2014) United Nations economic commission of Europe. http://www.unece.org/env/welcome.html. Accessed 26 June 2014

Velazquez L, Munguia N, Sanchez M (2005) Deterring sustainability in higher education institutions: an appraisal of the factors which influence sustainability in higher education institutions. Int J Sustain High Educ 6(4):383–391

Velazquez L, Munguia N, Platt A, Taddei J (2006) Sustainable university: what can be the matter? J Clean Prod 14(9–11):810–819

Support for Improved Scrap Tire Re-use and Recycling Decisions

Matthias Kalverkamp and Alexandra Pehlken

1 Introduction

Scrap tires have been with us since the development of the automobile and since their inception, there has been the problem of re-use or recovery of tire material. The composition of tires is, however, getting more complex which challenges recyclers to make new decisions regarding re-use or recycling. Some material flows are suitable to replace primary resources without loss of quality. Some materials are made never to separate by themselves and therefore pure material flows are impossible to achieve. A tool that considers different properties of material flows helps to evaluate the global recycling potential. Material Flow Analysis (MFA) and Life Cycle Assessment (LCA) can support the determination of the future potential of waste streams entering recycling processes. MFA can be easily applied to recycling processes because it takes into account all material flows entering and leaving the recycling process (system border). Since the input in recycling processes is often a mixture of various material streams, the exact composition is never known. There is often a lack of information due to unknown parameters in material composition or processing steps (Pehlken and Mueller 2009). Because of the high potential of recycling processes contributing to the sustainable management of resources (such as energy savings and material efficiency) it is necessary to assess the environmental impact of the material flows (Bringezu and Bleischwitz 2009).

M. Kalverkamp (✉) • A. Pehlken
Cascade Use, Carl von Ossietzky Universität Oldenburg, Oldenburg, Germany
e-mail: matthias.kalverkamp@uni-oldenburg.de; alexandra.pehlken@uni-oldenburg.de

J. Marx Gómez, B. Scholtz (eds.), *Information Technology in Environmental Engineering*, Springer Proceedings in Business and Economics,
DOI 10.1007/978-3-319-25153-0_4

In general, the following conditions for assessing sustainable resource management in recycling processes have to be achieved (Pehlken and Thoben 2011): There must be adequate material mass for:

(a) The recycling process and for
(b) Further product manufacturing
(c) Defined material properties and
(d) Very little variation of these properties

It is desirable to forecast a strategy for selecting recycling solutions that best fit the above-mentioned conditions with the goal of achieving a better environmental performance. A combination of the methods of MFA and LCA can help to roughly assess material flow, costs and environmental impacts. However, these methods lack the description of material properties, their variations and uncertainties respectively.

Scrap tires, as the case study in this paper, are a valuable source of secondary resources as they provide material flows of rubber, steel and fabric. There is potential for them to serve as secondary resource for new applications. However, the number of uncertainties in scrap tire recycling is numerous since only the output can be determined.

Steel can be easily recycled and is a valuable source of sustainable resource management because it replaces primary resources. Rubber is the biggest material stream derived from scrap tire processing. Rubber derived from scrap tires is a mixture of different qualities of rubber and added chemicals. Therefore, this material flow cannot compete with primary material for new tires, but this does not mean that the recycling of rubber means down-cycling and represents products of lower quality than the previous product.

The management of the life cycle of a product includes the most recent knowledge of processes and limits of systems. All recycling processes have in common that their input material has already had a complete life cycle; therefore, it is more difficult to find exact input data for the LCA of the recycling process. Uncertainties in performing LCAs in recycling processes must be taken into account.

The uncertainties in the recycling processes are the known and unknown parameters which have to be considered. Uncertainties can have many sources and Heijungs and Huijbregts (2004) state that the outcome of an LCA is affected by various types of uncertainty, such as parameters, scenario and model uncertainty. This uncertainty by itself has to be addressed in three places: the input side, the processing side and the output side. The variation of parameters is not always known and knowledge may be coincidental. For example, there are data for which no value is available, data for which an inappropriate value is available, and data for which more than one value is available. Residues always vary in their composition and only data ranges about material flow can be used as an input parameter. Notten and Petrie (2003) substantiate the statement that "different sources of uncertainty require different methods for their assessment". Due to different sources of uncertainty, this model relies not only on the quality of the process data but also on the uncertainty assessment.

2 State of the Art

2.1 Waste Management in the Automotive Industry

The European Parliament Council specifies quotas of material recycling in DIRECTIVE 2000/53/EC for End-Of-Life Vehicles (ELV) (European Parliament 2000). This so called ELV Directive established goals to minimise the effect of ELVs by setting recycling, reuse and recovery targets for the materials used in all manufactured vehicles. The directive requires that 95 % of ELV waste must be reused or recycled by 2015, with only 10 % of this recovered through the creation of energy. In addition to the pure recycling of materials, the re-use of components and sub-assemblies makes sense not only in line with the Lifecycle Management and Waste Act (Bundestag 2012) from Germany but also from the economic perspective due to limited resources, which are direct (material) and indirect (e.g. energy) resources (Östlin et al. 2009). This is also in accordance with the "Raw Material Initiative" released in 2008 by the European Commission (European Commission 2008). There is great pressure on saving and recovering raw materials in Europe.

In Germany, §4 of the Lifecycle Management and Waste Act clearly indicates that dumping of all kinds of waste materials has to be avoided before recycling procedures are undertaken (3R Hierarchy). According to the waste management (WM) hierarchy, reduction is the key factor for all WM strategies. By reducing the waste, manufacturers, users or societies have smaller material flows to recycle (European Commission 2008; Thierry et al. 1995). Re-manufactured parts are reused for their original purpose and perform the best when compared to most recycling processes. Re-manufactured products underline the market potential of the circular economy (Ellen MacArthur Foundation 2014). In the circular economy, recycling also has a high potential as a source for secondary resources; e.g. about one third of a VW Golf VII consists of recycled materials (Gernuks 2013). The use of secondary resources gains importance especially when combined with resource efficiency. Energy recovery remains the last option in cascade utilisation and disposal is the final handling of inert materials that are of no other use.

However, the focus of this research lies on re-use and recycling. The "Altfahrzeug-Verordnung (AltfahrzeugV)" (Bundesregierung 2013) regulates that all car manufacturers in Germany are obliged to take back their scrap cars and recycle them in dismantling facilities, following the ELV Directive. The following section describes a case study of scrap tires used in recycling.

2.2 Cascade Use of Scrap Tires as a Case Study

A good example of resource recovery from car parts is given by the potential of motor catalysts: The ELV directive has set clear rules for the handling of automotive catalysts in the EU: In 2007, 28 t platinum and 31 t palladium were recovered

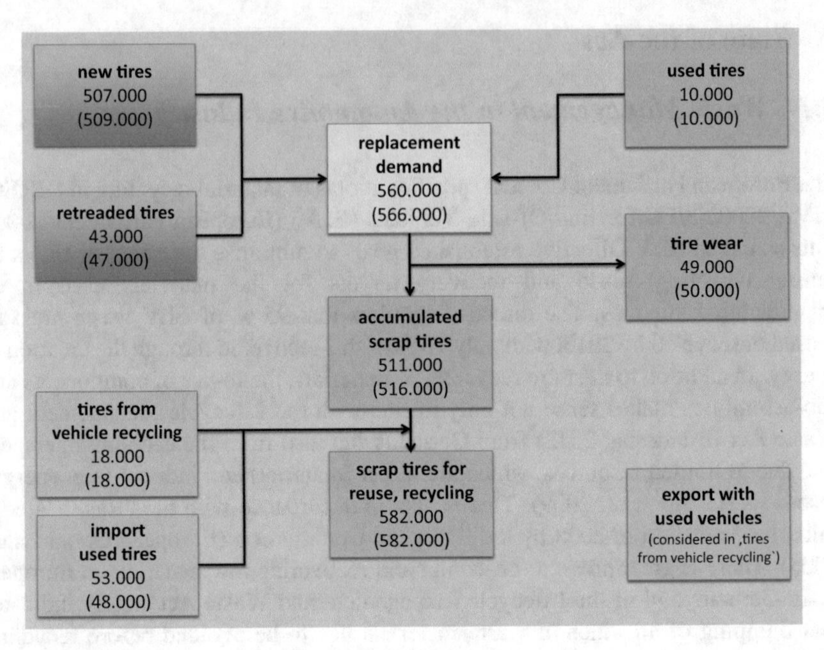

Fig. 1 Scrap tire usage in Germany in 2013 (WDK 2014)

from automotive catalysts globally (almost 15 % of the global mining production). However, despite this development it has to be underlined that even in the EU (especially in the new member states) collection systems are barely perfect. In contrast, the collection and recycling of scrap tires has been on the market since cars were first manufactured. Farmers are either using them in their storage systems or recyclers recover the rubber material, since natural and synthetic rubber possesses a good market potential for further treatment. Due to different market systems for dealing in scrap tires (public vs. private market structures), obtaining comparative figures for scrap tire recycling in the EU is difficult. These differences exist because the translation of EU regulations into national law is different. The focus therefore is on available numbers from Germany, a liberal market for scrap tire recycling, to describe the market situation. Figure 1 shows the statistics in Germany for scrap tire usage in 2013; the numbers in brackets indicate the previous year.

The scrap tire is what remains at the end of the life cycle of a tire. Some tires are re-treaded but usually, scrap tires are recycled into further applications. In Europe, more than 95 % of scrap tires are recycled. Used tires amounted to 3.3 million tonnes in the European Union countries in 2011. While energy recovery maintains its level around 38 %, material recycling is the real market benefiting from the policy of landfill diverting (ETRMA 2011). Energy recovery and material recycling represent 78 % of the whole scrap tire usage and by referring to the annual number of scrap tires nearly 3.1 million tonnes of material was being reused, either as material, energy or as a recycled tire itself. In the case of energy recovery, the tires replace fossil fuels and also by material recycling, a saving of primary resources, such as natural rubber occurs.

Although not a work of art, every scrap tire is in a sense unique, since each differs in its composition and age. Passenger tires are divided into summer, winter, high speed tires, etc. Passenger tires tend to contain more synthetic rubber than natural rubber when compared to truck tires. Truck tires consist of more natural rubber; off-the-road (OTR) tires, including heavy mining tires as well as agricultural and industrial tires, have nearly no synthetic rubber. The rubber composition may be because passenger tires have to meet higher quality standards (low rolling resistance, improved skid resistance and good wear) (Dunn and Jones 1991) to succeed in the competitive market. Truck and OTR tires, on the other hand, have to cope more with heavy loads and longer distances than with high speeds. The fibre content in passenger tires can be as much as 5 % of the total tire weight, whereas OTR tires tend to have little or no fibre content and contain about 15 % steel.

It is estimated that European tires have to be replaced within 3.5 years or 30,000 km and 6 years and 80,000 km depending on the mechanical load. The consequence is an uncertainty in the life expectancy of a tire and the exact date when the tire becomes a scrap tire is only an estimation. Recipes or formulae of the composition of tires are changing over time and therefore only mean values can be used for a LCA.

Although the exact composition of materials is not known in scrap tire recycling, recyclers have been very innovative and have adapted their processes to more or less guarantee quality specifications mainly for rubber, as the main output of the recycling plant. This rubber is either blended with plastics (rubber-plastic compound) or is used as single material to serve as input into new products, such as artificial turf, for example. The rubber-plastic blend produces even more innovative products and these are sometimes hard to identify as being recycled. Many of these products can be found in the interior of a car, for example, the dashboard or the door insulation.

2.3 Life Cycle Assessment and Material Flow Analysis

Life Cycle Assessment (LCA) is a method of evaluating the environmental impact of a product throughout its life, starting from the source, through the production and the usage phase till disposal—and potential iterations. LCA also provides meaning for economic evaluation because an LCA is able to elicit and bring to the surface "information about 'externalities" (Duda and Shaw 1997). The data from production processes is the basis for the evaluation. The data are usually an average value and not exact numbers for a particular product (Thorn et al. 2011). The procedures explained by ISO 14040/14044 are widely used and accepted to conduct an LCA. ISO 14040 defines the four major steps of an LCA: goal and scope definition, life cycle inventory analysis (LCI), life cycle impact analysis (LCIA) and life cycle interpretation. The first three steps are consecutive, although they can interface in both directions, whereas the last step, interpretation, refers to the three earlier steps and concludes the analysis (Thorn et al. 2011).

Besides LCA, material flow analysis is another tool to analyse environmental impacts and to improve resource management. “Material flow analysis (MFA) is a systematic assessment of the flows and stocks of materials within a system defined in space and time.” (Brunner and Rechberger 2004). Materials in MFA are both goods and substances, where goods consist of different substances (Cencic and Rechberger 2008). Typical systems observed in this context are factories or households. The relevant inputs and outputs in these have to be defined according to the defined space Material flows within such systems are followed from their source until their end, passing through different process steps (transport, transformation, storage) until reaching their final conclusion. By developing alternative scenarios for a system, the initial scenario can be evaluated against these alternatives, thus supporting system design improvements, changes and adoptions (Brunner and Rechberger 2004).

In the case of scrap tires, LCAs are already widely used to verify most favourable treatment options for the environment. Traditionally, scrap tires are used in cement kilns as a fuel replacement, but they are used as a secondary material for ground rubber and in civil engineering applications (Feraldi et al. 2013; Li et al. 2014). However, incineration is the least preferable WM strategy—before disposal/landfill. Feraldi et al. (2013) investigate in their study which treatment option for scrap tires has the least impact on the environment, material recycling or incineration, i.e. energy recovery. They found evidence that, material recycling has a less harmful impact on the environment than incineration. More detailed studies show how the application of technologies to scrap tire recycling processes and the production of renewable energy can reduce the environmental impact of scrap tire recycling (Li et al. 2014).

The before-mentioned results from existing LCAs demonstrate how the method already contributes to a better understanding and evaluation of scrap tire treatment options. The proposed decision-support tool aims to incorporate the existing methods, together with additional knowledge, about the recyclability to improve decision making further when WM for scrap tires is considered.

3 Method and Concept for Cascade Use of Products

3.1 Decision Support

The process for WM decisions will be supported by a software tool. It is expected that such a tool will help to lower the level of complexity for the user, the person who makes the decision. In the case of scrap tires, this could be car dismantlers, tire dealers or people in other facilities for receiving scrap tires. The decision room targeted by the decision tool is illustrated in Fig. 2. It follows the product life cycle

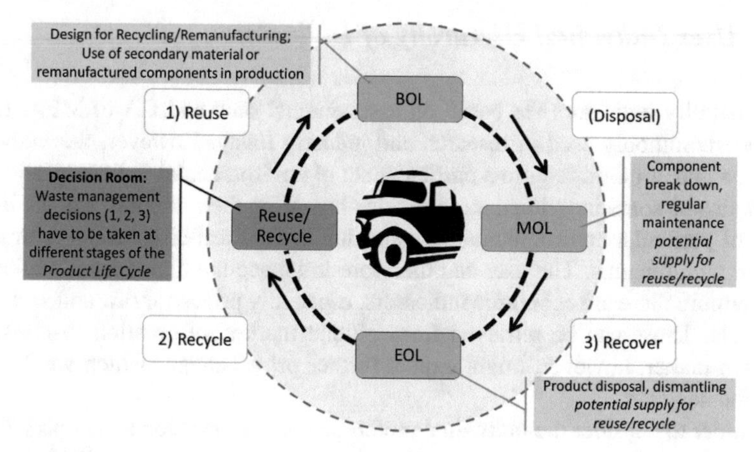

Fig. 2 Decision room between re-use and recycle at the end-of-life of a vehicle

separated into begin-of-life (BOL), mid-of-life (MOL) and end-of-life (EOL). The following order of decisions in scrap tire waste management is ecologically preferable: re-use, recycle and recover. The disposal of scrap tires is banned in the EU.

The decision tool will be developed in a user-oriented way. The users will define and choose their own sustainability criteria with either a stronger focus on ecological or economic criteria. Support for the raw material assessment will be given by algorithms for information on material flow and the sustainability assessment implies that the parameters for material will be defined. Therefore, it is necessary that the model is designed in a manner which can be adapted for various material flows and recycling processes.

The decision tool has to cope with highly dynamic situations in the recycling market, due to price volatility at in market and changing technologies for material separation or re-use. For example, it is necessary to consult stock exchanges to get market information on recycling. There are already data bases available that might be useful for this software (ecoinvent,[1] Probas,[2] etc.). In addition, European researchers are developing new catalogues (ELCD/ILCD) based on standard processes for key materials, energy carriers, transport and WM. These databases are all taken into account during the development of the decision tool. Besides the main decision tool, the development of an app for mobile devices is planned to directly visualise the environmental impact of a WM decision for a set of particular scrap tires according to sustainable indicators.

[1] http://www.ecoinvent.org/database/

[2] http://www.probas.umweltbundesamt.de

3.2 *User Individual Hierarchy of Indicators*

Sustainability indicators are based on assessments, such as LCA or MFA. Hence they are commonly used in research and industry today. However, the individual decision maker might require a particular set of environmental indicators and there might also be some individual requirements regarding their ranking, e.g. the importance of particular environmental impacts due to the local circumstances or information requirements. The user can therefore influence the hierarchy of indicators. Furthermore there are economic indicators, especially prices for raw and secondary materials. Those can be retrieved from official market information systems. The decision maker, however, might expect further price changes which would influence the decision.

In order to consider the individual preferences of the decision maker properly, in combination with the general knowledge about tire composition, available technologies for resource recovery and the actual market situation, the decision support tool uses an Analytic-Hierarchy-Process (AHP) based approach. The AHP is a known approach in multi-criteria decision analysis. The AHP was chosen due its capability of dealing with complex decisions, although it is also rather complex in its application. Nevertheless, especially in comparison with other methods, the AHP is less arbitrary (Rosenkranz 2006; Saaty 2008).

4 Proposed Software Tool

One of the challenges in WM is to make the right decision regarding environmental impact. However, decisions about the WM option taken by a company are mainly driven by economic considerations, due to profit orientation. This makes decisions more complicated For example, the whole process from acquisition (procurement of EOL products) until final sales of a product (reuse), as a secondary material (recycle) or as a fuel (recovery) has to be cost efficient. Therefore, environmentally better treatment options which would be better for the environment might be overruled by economic considerations. However, the environmentally better option might still not be a losing deal-only the resulting benefit might not be so profitable. However, the economically preferred solution could also be ecologically a good solution. When adding legal questions and regulations to this decision situation, the complexity increases further.

Hence, a software tool is proposed that is able to consolidate the different aspects relevant for decisions in WM. There are laws and regulations for treatment options (e.g. quotas), markets especially for different sales options and environmental assessment representing special societal requirements (technology e.g. regarding changes in the processing of secondary material). While consolidating this information, the tool will not take the decision on its own but will rather support the decision maker in the decision process. For example, the tool will be able to suggest alternative WM scenarios due to the changing market prices for particular secondary materials. This allows the decision maker to revise the decision. This revision is useful and necessary, because the decision maker might also have additional information about potential demand or other market options, e.g. due to technological developments, not yet known to the software. Therefore, it is proposed that the tool is a supporting solution but not a substitution. Since the decision maker cannot be omniscient about all economic factors (Richter and Furubotn 2010) when making a decision, the software tool will increase transparency and lower transaction costs for decision-making processes about waste treatment. Eventually, the tool will help to balance ecological and economic interest in waste management in order to increase sustainable decisions.

The software is based on different inputs at different levels, as depicted in Fig. 3. First, there is a level of general indicators, criteria and data, which are provided by different sources. For example, LCAs provide environment indicators and regulations that define criteria for waste and other substances. This information is fed into the software tool for further processing. Second, there is a level where the user can adjust criteria and indicators according to individual requirements and also individual knowledge. Finally, the software itself processes these two inputs, data and the individualised hierarchy of a user, with the help of an analytical hierarchy process.

As mentioned above, the user can take another decision based on his/her own judgement., independently from the tool's result regarding the decision If this information is fed back into the tool (i.e. selection button), the tool will be able to take this decision into account when it comes to the next decision. The reasons for

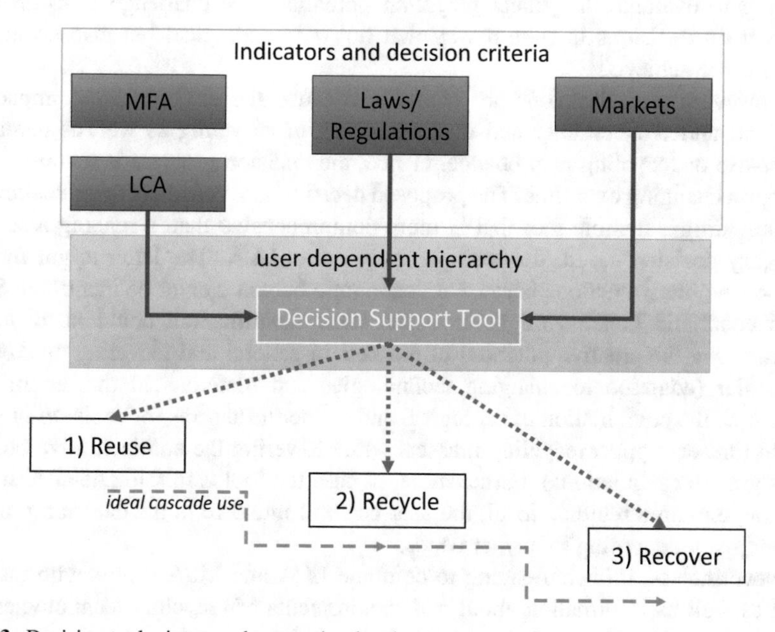

Fig. 3 Decision tool—input and processing levels

this are (a) to reconstruct the decision, (b) to allow for forecasts regarding and compliance with the legal quotas, and (c) to facilitate reporting afterwards.

The advantages of the proposed software tool are its complex and different data sources, such as:

- Databases from environmental assessment
- Databases regarding the composition of car parts
- Price information, e.g. market information from a commodity exchange
- Demand information in terms of particular demands (e.g. from nearby production facilities)
- Laws and regulations regarding the WM (e.g. prohibited or other treatment options, treatment quotas)

5 Discussion and Outlook

Sustainable management of resources together with an integrated policy of re-use and recycling strategies are necessary in order to address environmental challenges. While some material flows are appropriate to replace primary resources, other materials are only useful for products of minor quality. Some materials are made never to separate by themselves and therefore pure material flows are impossible to achieve. A tool that considers the properties of different materials and the supply of materials and the surrounding environmental and ecological conditions is essential to help to evaluate the global recycling potential. The challenge is to provide support on decisions in such a way that not only ecological but also economic benefits are achieved.

However, such decisions are complex because the environmental impact is assessed under uncertainty and the technology of recycling as well as demands for re-use or recycling may change. Hence, the decision problem is manifold and keeps on changing over time. The proposed decision tool considers these challenges by providing a flexible tool that is more comprehensive than a making a single-category decision based, for example, only on an LCA. The latter might further support regulative actions because an economic benefit cannot be identified. Still, such economic benefits might exist but their identification could be difficult. Considering the creative potential of markets in general and recycling markets in particular (adaption to changing technologies and materials in the automotive industry), the combination of ecological and economic aspects for decision support could further support recycling markets while lowering the administrative burden (less granular regulations). Furthermore, in case the tool is missing detailed information, e.g. on a regional level, the user can still interfere in the decision process and adjust it according to his/her needs.

Nevertheless, it is challenging to combine LCA and MFA results with market data as well as information about and requirements of recycling technologies. In addition, the tool might only be suitable for specific car parts and not for the whole car. Some parts are still not investigated for reuse and the only option is material or energy recovery.

Acknowledgments The research for this paper was financially supported by the German Federal Ministry of Education and Research (Grant no.: 01LN1310A). The authors gratefully acknowledge their support.

References

Bringezu S, Bleischwitz R (2009) Sustainable resource management: global trends, visions and policies. Greenleaf, Sheffield

Brunner PH, Rechberger H (2004) Practical handbook of material flow analysis, advanced methods in resource and waste management. Lewis, New York

Bundesregierung (2013) AltfahrzeugV—Verordnung über die Überlassung, Rücknahme und umweltverträgliche Entsorgung von Altfahrzeugen. https://www.juris.de/purl/gesetze/AltautoV. Accessed 15 Jan 2015

Bundestag (2012) KrWG—Gesetz zur Förderung der Kreislaufwirtschaft und Sicherung der umweltverträglichen Bewirtschaftung von Abfällen. https://www.juris.de/purl/gesetze/_ivz/KrWG. Accessed 15 Jan 2015

Cencic O, Rechberger H (2008) Material flow analysis with software STAN. In: Möller A, Page B, Schreiber M (eds) EnviroInfo 2008: environmental informatics and industrial ecology. 22nd international conference on informatics for environmental protection, Leuphana University, Lüneburg, Sept 2008. Shaker, Aachen, pp 440–447

Duda M, Shaw JS (1997) Life cycle assessment. Society 35(1):38–43

Dunn JR, Jones RH (1991) Automobile and truck tyres adapt to increasingly stringent requirements. Elastomerics 123(7):11–18

Ellen MacArthur Foundation (2014) Towards the circular economy: accelerating the scale-up across global supply chains. http://www.ellenmacarthurfoundation.org/business/reports/ce2014#. Accessed 1 Feb 2015

ETRMA (2011) European tyre & rubber manufacturers' association. http://www.etrma.org/. Accessed 15 Jan 2015

European Commission (2008) Communication from the commission to the European Parliament and of the council. http://eur-lex.europa.eu/LexUriServ/LexUriServ.do?uri=COM:2008:0699:FIN:en:PDF. Accessed 20 Jan 2015

European Parliament, Council of European Union (2000) DIRECTIVE 2000/53/EC of the European Parliament and of the Council of 18 September 2000 on end-of life vehicles. Off J Eur Communities 43(L 269):34–42

Feraldi R, Cashman S, Huff M, Lars R (2013) Comparative LCA of treatment options for US scrap tyres: material recycling and tire-derived fuel combustion. Int J Life Cycle Assess 18 (3):613–625

Gernuks M (2013) Beitrag von Recycling zur Rohstoffversorgung am Beispiel Automobilindustrie. http://www.deutsche-rohstoffagentur.de/DERA/DE/Downloads/Gernuks_BGR-Rohstoffkonferenz2013.pdf. Accessed 30 July 2014

Heijungs R, Huijbregts MAJ (2004) A review of approaches to treat uncertainty in LCA. In: Pahl-Wostl C, Schmidt S, Jakeman T (eds) iEMSs 2004 international congress: "complexity and integrated resources management". International Environmental Modelling and Software Society, Osnabruck, June 2004. Complexity and integrated resources management—transactions of the 2nd Biennial meeting of the international environmental modelling and software society, vol 1. International Environmental Modelling and Software Society, Manno, pp 332–339

Li W, Wang Q, Jin J, Li S (2014) A life cycle assessment case study of ground rubber production from scrap tyres. Int J Life Cycle Assess 19(11):1833–1842

Notten P, Petrie J (2003) An integrated approach to uncertainty assessment in LCA. In: Proceedings of the international workshop on Life Cycle Inventory (LCI) Data, Institut für Technishe Chemie, Hannover, 20–21 Oct 2003

Östlin J, Sundin E, Björkman M (2009) Product life-cycle implications for remanufacturing strategies. J Clean Prod 17(11):999–1009

Pehlken A, Mueller DH (2009) Using information of the separation process of recycling scrap tyres for process modelling. Resour Conserv Recycl 54(2):140–148

Pehlken A, Thoben K-D (2011) Contribution of recycling processes to sustainable resource management. In: Proceedings of the 18th CIRP international conference in life cycle engineering, Technische Universität Braunschweig, Braunschweig, 2–4 May 2011

Richter R, Furubotn EG (2010) Neue Institutionenökonomik. Mohr Siebeck, Tübingen

Rosenkranz F (2006) Geschäftsprozesse: Modell- und computergestützte Planung, 2nd edn. Springer, Heidelberg

Saaty TL (2008) Decision making with the analytic hierarchy process. Int J Serv Sci 1(1):83–98

Thierry MC, Salomon M, van Nunen JAEE, van Wassenhove LN (1995) Strategic issues in product recovery management. Calif Manag Rev 37(2):114–135

Thorn MJ, Kraus JL, Parker DR (2011) Life-cycle assessment as a sustainability management tool: strengths, weaknesses, and other considerations. Environ Qual Manag 20(3):1–10

WDK (2014) Altreifenmenge konstant. http://news.wdk.de/de/Pressemitteilung.html?articleID=16018. Accessed 15 Jan 2015

Risk Profiling for Corporate Environmental Compliance Management

Heiko Thimm

1 Introduction

Corporate environmental compliance management is a work area where specialists from multiple different disciplines such as occupational safety, hazardous material management, fire protection, and transportation safety inherently need to work together. The compliance management team together with collaborators from other work areas is obligated to assure that the company complies with all relevant regulations. Activities that are required in order to enforce environmental compliance can be oriented at many aspects of business organizations including product properties, production and logistic processes, corporate infrastructure, and the workforce's level of competencies and skills (Thimm 2015). The company faces sanctions ranging from fines, withdrawal of licenses and even permits to mandatory closures and shutdowns when full compliance is not maintained at all times (Gunningham 2011). The enforcement efforts that are required in order to assure environmental compliance can be disturbed and even aborted by uncertain events. Considering general risk management paradigms and best practices such as given in the ISO 3100 standard (ISO 2009) it can be recommended to solve this problem—i.e. "the effect of uncertainty on objectives"—through an extension of corporate compliance management by risk control aspects (IMA 1995). Supporting evidence for this recommendation can be found in the body of international environmental policies and governmental practice that are traditionally promoting environmental risk control aspects (Gunningham 2011; IMPEL 2012; HM Treasury 2004).

In the light of the enormous amount of regulations and the severe potential sanctions two surprising facts can be observed about today's corporate environmental compliance practice. First, there still exist companies that show a limited

H. Thimm (✉)
School of Engineering, Pforzheim University, Pforzheim, Germany
e-mail: heiko.thimm@hs-pforzheim.de

J. Marx Gómez, B. Scholtz (eds.), *Information Technology in Environmental Engineering*, Springer Proceedings in Business and Economics,
DOI 10.1007/978-3-319-25153-0_5

awareness of the risk of non-compliance and therefore do not address this risk systematically (Walker et al. 2008). Second, the compliance management tasks are often completed with assistance of rudimentary IT solutions such as spread sheet software (McKeiver and Gadenne 2005; Walker et al. 2008) that bear many inherent problems like a low efficiency and a high probability for data inconsistencies.

Our research aims on the development of a novel data provisioning service for environmental compliance management information systems (ECM IS) (Freundlieb and Teuteberg 2009; McKeiver and Gadenne 2005). The need for such as new service is motivated by empirical data. We refer to this service by "risk profiling service" because the users are through the service continuously informed about the risk of non-compliance. In this aspect our research contributes to a new environmental compliance enforcement approach promoted mainly by the European Union Network for the Implementation and Enforcement of Environmental Law (IMPEL) (IMPEL 2012). On the basis of a specialized supervision method IMPEL promotes the use of suitable tools such as the proposed ECM IS in order to enable companies to pro-actively self-control environmental risks. The risk conceptualization proposed in this article is built upon the fact that successful corporate compliance enforcement largely depends on carefully chosen and managed human-lead enforcement activities (Freundlieb and Teuteberg 2009; Gunningham 2011). It is addressed that these activities bear many risks (e.g. wrong decisions, incorrect judgments, illness of persons, failures in compliance enforcement plans) that can among others result from inherent phenomena of the work of single persons and of group work. It is expected that through the service problems with compliance enforcement activities can be detected on an early stage. By reacting to these problems appropriately it is possible to prevent future non-compliance states.

Following this introduction in Sect. 2 an overview of related research work is given. The risk management framework is described in Sect. 3. Central considerations for the extension of ECM IS by the proposed risk profiling service can be found in Sect. 4. Concluding remarks are contained in Sect. 5.

2 Related Works

Some of the principles proposed in this article to improve corporate environmental compliance are related to the reference model for Environmental Management IS by Freundlieb and Teuteberg (2009). However, their data warehousing based reference model is to a large extent focused on providing reporting and analysis capabilities without addressing risk issues.

Several research groups investigated the use of database technology for corporate compliance management primarily focusing on financial compliance. Researchers from IBM and the University of Stuttgart proposed a system where similar to our approach compliance management activities are logged in database tables in order to detect anomalies that may indicate compliance violations (Agrawal et al. 2006). A general discussion of methods to maintain tamper-resistant audit logs for legal compliance purposes can be found in (Snodgrass et al. 2004).

A research group at the Stanford University investigated the use of Internet technologies, text mining, and information retrieval techniques in order to provide improved public services for environmental compliance management tasks (Kerrigan 2003). Despite the fact that instead of public web-based support our research is focused on support of corporate compliance experts there exist several communalities that are noteworthy. In particular, the Stanford group's Regulation Assistance System targets to assist the users in compliance checking on the basis of a XML regulation framework. The regulations and compliance rules are represented as XML documents that are tagged with logic that can be exploited for checking compliance. The proposed web-based system takes as input documents from the user and possibly further interactive data, checks the data for contradictions, and makes a compliance decision. While a similar objective is addressed by the proposed risk profiling service the checking approach is different and focused on irregularities given in data about compliance enforcement activities.

The foundation of the research described in this article has been reported in an earlier paper (Thimm 2015) that contributed a best business practice for a process-oriented development, implementation and use of IT for the assurance of environmental law compliance in Small and Medium Sized Enterprises (SME). Based on a proposed reference model an initial version of the new ECM IS solution, Compliance Center Professional (CCPro), is being developed. The data model and major concepts of the CCPro system are described. The extension of these earlier contributions by the consideration of the issue of risk of non-compliance, however, is a recent progress in our long term research program and, thus, not addressed in the earlier article.

3 A Risk Management Framework for Corporate Environmental Compliance

The risk management framework described in the following is largely drawn from earlier empirical investigations of the environmental compliance work area (Thimm 2015). The main informant has been a manufacturer of chemicals, industrial alcohol, and plastics situated in South Germany which belongs to the category of Small and Medium Size Enterprises (SME). For more than 100 years the company has been successful within its growing international market. About 4 years ago the company started to successfully offer environmental management consulting services to other companies. Through the large number of completed internal and external projects and through a lot of activities in corresponding expert networks deep knowledge could be obtained and contributed to the research described in this article. The research team was especially provided with knowledge

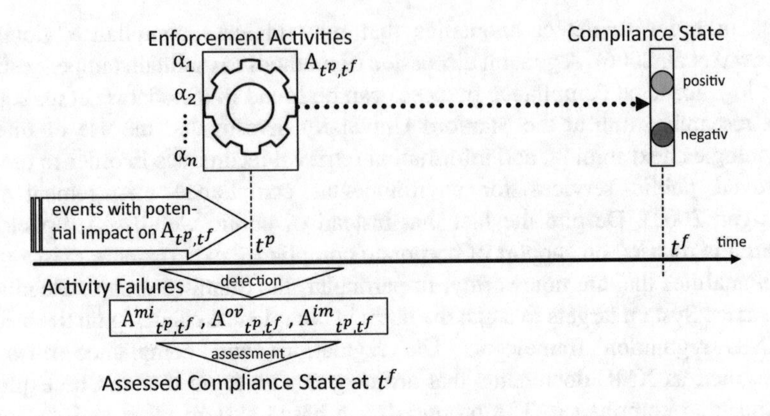

Fig. 1 Compliance state estimation from enforcement activity failures

about typical compliance management work processes and potential reasons causing compliance violations. The empirical findings were complemented by information from a literature study that was focused on guidelines and experience reports published in journals for environmental professionals.

In order to permanently assure that the set of all relevant environmental compliance regulations are fulfilled companies need to perform a multitude of compliance enforcement activities (Gunningham 2011; IMA 1995; Thimm 2015). In Fig. 1 the box with the gear-wheel symbol labelled A_{t^p,t^f} represents the subset of at the present time point t^P ongoing compliance enforcement activities referred to by α_i $(i = 1, 2, \ldots, n)$. The activities α_i are directed at the compliance state at time point t^f. In general, they can be categorized into the following different activity types: (1) activities to monitor announcements of new regulations and of amendments of existing regulations, (2) activities to determine the relevance of new regulations and of revisions of existing regulations, (3) activities to determine measures that are required in order to comply with regulations, (4) activities to implement measures that have been determined, (5) activities to check the effectiveness of measures, and (6) documentation activities.

The measures can be oriented at the typical environmental health subjects such as products, production processes, customers, employees, and infrastructure components. Similar to other business activities also compliance enforcement activities are required to be performed in a well synchronized fashion such that the obtained results meet given conditions (IMPEL 2012). Therefore, in the proposed framework enforcement activities are treated as obligations with associated deadlines and outcome constraints. Obviously, here exists a positive correlation between a company's environmental compliance status and the company's compliance enforcement activities. When at a present point in time t^P the activities α_i $(i = 1, 2, \ldots, n)$ of A_{t^p,t^f} are neither violating their deadline nor their respective outcome constraints then at a corresponding future time point t^f compliance will be achieved. Conversely, if an activity exists that is overdue or/and that failed to satisfy all outcome constraints a non-compliance status has to be assumed for t^f.

One has to admit the rule that it is not possible to guarantee that enforcement activities are always completed in time and always fully meet their outcome constraints. Among others two simple reasons for this rule are (1) that many of the environmental compliance activities involve human executants and (2) that activities intended to implement compliance assurance measures may suffer from material and equipment defects. This uncertainty motivates to extent corporate environmental compliance management by awareness of the risk of non-compliance due to problems with enforcement activities. In order to address this awareness two specific steps are defined which are completed at time point t^p in order to assess the compliance state for time point t^f. The first step attempts to detect the activity failures. This can be performed, for example, by the use of business process mining and prediction techniques (de Leoni et al. 2014) in combination with data analysis methods targeted on "hard indicators" and "soft indicators" (Bullen et al. 2009). The failure detection is followed by a second step in which the impact of the failures on the compliance state at time point t^f is assessed. At the present state of our research we are considering the following three different types of activity failures

By *missed activities* it is referred to a set of activities $A^{mi}{}_{t^p,t^f}$ that in spite of being relevant for the compliance state at time point t^f do not belong to the compliance enforcement activities at time t^p (i.e. $A^{mi}{}_{t^p,t^f} \cap A_{t^p,t^f} = \{\}$). A missed activity, for example, can refer to a situation where the compliance team overlooked the announcement of a new community regulation for noise immission that will become effective at a certain future date. Consequently, the new regulation will not be addressed and cause a future non-compliance state. Checking "hard indicators" in order to detect missed activities means to compare the usual pattern of monitoring activities with the most recent pattern. Deviations between the patterns are to be treated as potential activity failures. Checking "soft indicators" can be performed by evaluating external information sources that store regulation announcements such as online information services and regulation portals. For example, assume that one identified a time period in which many new regulation announcements were published. When during this time period only a small number of monitoring activities is found in the compliance management data it can be assumed that the company missed some monitoring activities.

By *overdue activities* it is referred to a set of activities $A^{ov}{}_{t^p,t^f}$ (with $A^{ov}{}_{t^p,t^f} \subseteq A_{t^p,t^f}$) that consists of the enforcement activities α_j $(j = 1, 2, \ldots, m;\ m \leq n)$. The activities α_j are either overdue at the present time point t^p or they will miss the deadline during the time interval $]t^p, t^f]$. To give an example assume that the above described noise regulation was recognized and in a next step evaluated to be relevant for the company. As further activity it will be required to decide about measures to fulfill the noise immission regulation. That is, one needs to schedule a corresponding decision activity and set an activity deadline. When the decision activity is not completed in time then it can happen that the measure implementation activity will not be completed in time, too. This activity failure can ultimately cause the company to fail the new noise immission regulation and result into a non-compliance state. In order to detect overdue activities one can simply check if activities with violated deadlines exist. In order to complement this approach by an analysis that is directed at "soft indicators" one can consider activities that have not violated their deadlines (yet) but that are likely to fail the given deadline. In order to identify such activities many aspects can be considered including the complexity of the activity and the resources required. For example, consider a complex measure implementation activity that involves many people. When the remaining time span to the deadline is very short it can be assumed that the activity will most likely fail the deadline.

By *imperfect activities* it is referred to an activity set $A^{im}{}_{t^p,t^f}$ (with $A^{im}{}_{t^p,t^f} \subseteq A_{t^p,t^f}$) which is composed of activities α_k ($k = 1, 2, \ldots, l$; $l \leq n$). The activities α_k correspond to either terminated activities that violated outcome constraints or ongoing activities that will terminate during the time interval $]t^p, t^f]$ with outcome constraint violations, too. As an example, consider again the scenario of the noise immission constraint. Assume that in a further step it was decided to implement a noise immission fence. An imperfect activity refers to the situation that the fence (i.e. measure being implemented through a corresponding implementation activity) will not well enough shield the noise immission. As a result of the failed outcome constraint the noise limit will be violated which leads to a non-compliance status. The detection of terminated activities that violated outcome constraints is a relatively simple checking task. The detection of constraint violations for still ongoing activities is a more complex task that can be performed, for example, by the use of appropriate prediction methods (Leitner et al. 2009; Salfner et al. 2010) and by expert judgements.

4 An IS-Based Risk Profiling Approach

Environmental compliance management information systems (ECM IS) inherently deal with information objects that are processed in the proposed risk management framework (Freundlieb and Teuteberg 2009; Thimm 2015). It therefore seems to be a natural consideration to extend ECM IS by a novel risk management service referred to in the following by risk profiling service. The service performs two steps that are described in the next section. This is followed by a section with an overview of a sample implementation approach.

4.1 Risk Estimation and Risk Aggregation

On the basis of the risk framework of Sect. 3 the further conceptualization work builds on general principles used in occupational safety, health risk assessment (Nunes 2013), and risk aggregation (David 2008; HM Treasury 2004). In order to express different levels of risks a simple five point qualitative scale is used with the values (i.e. risk categories) very low risk, low risk, medium risk, high risk, very high risk. Through data about the current set of enforcement activities future risk levels are estimated based on three factors. The probability that the risk will materialize in reality is considered as first factor which is assigned a qualitative scale with the values very unlikely, unlikely, likely, more likely, very likely. The level of severity by which the compliance status will be impacted when the risk materializes in reality serves as second factor (scale: very low, low, medium, high, very high). The third factor refers to the ability to recover a positive compliance status considering the resources and capabilities demanded and being available for a full recovery (scale: very good, good, satisfactory, limited, very limited).

To demonstrate the general principles of the estimation method three same scenarios are described next. As first scenario consider an overdue deadline for a compliance enforcement measure intended to instruct workers to place toxic waste material in a special waste disposal box. Under normal circumstances for a lot of companies the estimation method will assess a very low risk for this hazard. This result is implied by the fact that one can assume the severity of the impact on the compliance status to be very little. Furthermore, the ability to recover can be assumed to be very good. As second scenario let us assume that there exists the hazard of an overdue deadline for enforcement decisions that are directed at occupational safety infrastructure components. Examples for such measures to prevent accidents in the production plant are safety cages and fences to separate machines from walkways. It can be assumed for a lot of companies that under normal circumstances a medium or even high risk for this hazard will be assessed. To also describe a scenario of a very high risk of a negative compliance status consider as third example a company that develops and produces consumer products. When the engineering department of the company fails to replace a carcinogens part of a new product before the production start then the environmental compliance status of the company is at a high risk.

The above described risk estimation approach is focused on estimating a risk value for a given single risk instance. In a further step all single risk values are to be aggregated in a meaningful way so that users can see all risks at one glance. There exist a number of different risk aggregation approaches that according to David (2008) can be classified into different aggregation types. For our risk profiling service we use a risk aggregation approach that is targeted at showing the total exposure to loss which an organization faces across its portfolio of operations. Showing the aggregation result can be effectively performed through the generation of a risk profile that enumerates all risk types for which a risk value has been obtained. Per risk type for each of the risk categories the respective numbers of identified risks can be presented in the form of a bar chart.

4.2 Risk Profiling IS-Service: Sample Implementation Approach

The proposed risk profiling approach is currently being implemented as a new data provisioning service of the information system Compliance Center Professional (CCPro). CCPro is based on an earlier research prototype of an innovative ECM IS (Thimm 2015) that has been further developed and improved towards typical requirements of business information systems. CCPro is now available in a beta version that is provided to selected companies for evaluation and testing purposes.

Figure 2 shows a screen of a special version of CCPro where the main menu item "Risk Radar" has been selected. As a result compliance status information and also the risk profiles for three future time point are displayed in the action area of the window. This screen of CCPro has been devised as a design study in order to present our vision for compliance risk management to CCPro users and others. The overall structure of the risk profiling service being implemented as an extension of CCPro is shown in Fig. 3. The database of CCPro stores data about compliance regulations, announcements, and all ongoing and completed compliance management activities as well as measures. The Regulation Registry, which is a logical subset of the database, stores a company specific set of legal regulations which are valid for a specific time period and that are evaluated with respect to their relevance for the company. The Compliance Management Log which is a further logical subset of the database corresponds to what sometimes is referred to by Audit Log (Snodgrass et al. 2004). This log is used in order to record all planed and performed compliance enforcement activities together with the obtained activity results. The Risk Management Knowledge Base contains the set of hard and soft indicators which are the basis of the data analyses to identify activity failures. This component

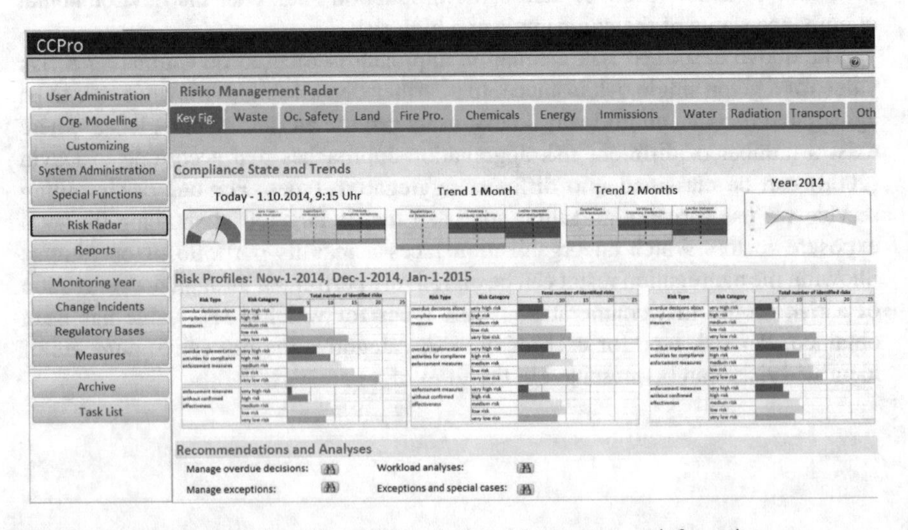

Fig. 2 CCPro screen showing risk profiles and other risk management information

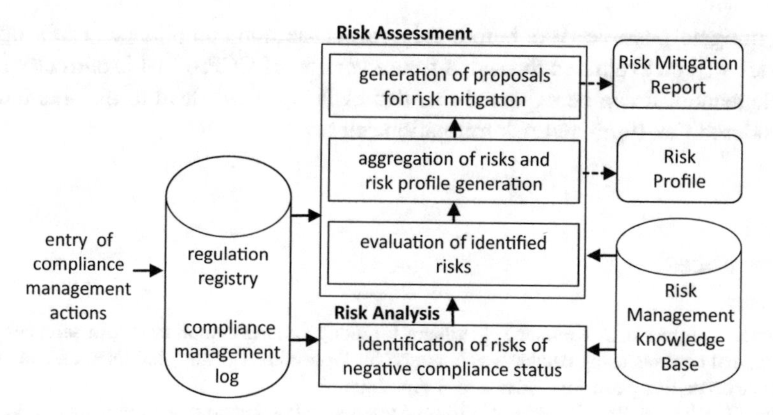

Fig. 3 Components of the proposed risk profiling service

also contains risk mitigation rules that are used to generate advice for the users on how to mitigate identified risks.

The data processing steps are grouped into a Risk Analysis part and a Risk Assessment part. By using the Risk Management Knowledge Base and through corresponding analytical methods targeted at the Regulation Registry and the Compliance Management Log the Risk Analysis first identifies relevant risks. The Risk Assessment is structured into three steps: (1) the risk evaluation step performs a qualitative risk evaluation according to the earlier described principles—that is the risks are estimated with respect to the three factors probability of risk, severity of harm, and ability for full recovery; (2) the risk aggregation step performs an aggregation of the set of quantified single risks and generates a corresponding risk profile in the form of a bar chart; (3) the third step generates proposals for risk mitigation actions through corresponding mitigation rules defined in the Risk Management Knowledge Base.

5 Conclusions

The world population's increasing awareness for the big environmental problems and the continuing trend for a global harmonization of corporate environmental standards created an enormous amount of legal regulations that are oriented at the different fields of environmental protection laws. The large amount of regulations and the threatening sanctions when a company does not comply with relevant laws imposes increasing pressure for companies to address the inherent risks of compliance management tasks. The proposed risk profiling information service can help companies to obtain an effective and risk aware environmental compliance practice. Through the frequent provision of the company's up-to-date risk profile the compliance management experts can get a good understanding of the risk situation and mitigate possible risks before they can cause non-compliance. The proposed service will be evaluated through a future version of CCPro that is currently being implemented. It can be expected that this evaluation will lead to insights into the considered risk types and risk mitigation rules.

References

Agrawal R, Johnson C, Kiernan J, Leymann F (2006) Taming compliance with sarbanes-oxley internal controls using database technology. In: Proceedings of the 22nd international conference on data engineering, Atlanta, 3–7 Apr 2006

Bullen G, Dickson RM, Roberts W (2009) Systems and/or methods for prediction and/or root cause analysis of events based on business activity monitoring related data. US patent 20090171879 A1

David SR (2008) Safety risk aggregation: the bigger picture. http://www.sars.org.uk/old-site-archive/forms/david.pdf. Accessed 15 June 2015

de Leoni M, van der Aalst WMP, Dees M (2014) A general framework for correlating business process characteristics. In: Sadiq S, Soffer P, Völzer H (eds) Business process management. 12th international conference on business process management, Haifa, Israel, Sept 2014. Lecture notes in computer science, vol 8659. Springer, Berlin, pp 250–266

Freundlieb M, Teuteberg F (2009) Towards a reference model of an environmental management information system for compliance management. In: Wohlgemuth V, Page B, Voigt K (eds) EnviroInfo 2009: environmental informatics and industrial environmental protection: concepts, methods and tools. 25th conference on environmental informatics and industrial environmental protection: concepts, methods and tools, Berlin, Sept 2009. Shaker Verlag, Aachen, pp 129–138

Gunningham N (2011) Enforcing environmental regulation. J Environ Law 23(2):169–201

HM Treasury (2004) The orange book management of risk—principles and concepts. https://www.gov.uk/government/uploads/system/uploads/attachment_data/file/220647/orange_book.pdf. Accessed 15 June 2015

IMA (1995) implementing corporate environmental strategies, business performance management 67. Institute of Management Accountants (IMA), Montvale, New Jersey. http://www.imanet.org/docs/default-source/thought_leadership/management_control_systems/implementing_corporate_environmental_strategies.pdf?sfvrsn=2. Accessed 15 June 2015

IMPEL (2012) Compliance assurance through company compliance management systems. European Union network for the implementation and enforcement of environmental law (IMPEL). http://impel.eu/projects/compliance-assurance-and-company-compliance-management-systems/. Accessed 15 June 2015

ISO (2009) ISO 31000:2009—risk management. International Standards Organization (ISO)

Kerrigan SL (2003) A software infrastructure for regulatory information management and compliance assistance. Dissertation, Stanford University

Leitner P, Wetzstein B, Rosenberg F, Michlmayr A, Dustdar S, Leymann F (2009) Runtime prediction of service level agreement violations for composite services. In: Dan A, Gittler F, Toumani F (eds) Service-oriented computing. 7th international joint conference on service oriented computing, Stockholm, Nov 2009. Lecture notes in computer science, vol 6275. Springer, Heidelberg, pp 176–186

McKeiver C, Gadenne D (2005) Environmental management systems in small and medium businesses. Int Small Bus J 23(5):513–537

Nunes I (2013) Occupational safety and health risk assessment methodologies. http://oshwiki.eu/wiki/Occupational_safety_and_health_risk_assessment_methodologies. Accessed 04 Jan 2015

Salfner F, Lenk M, Malek M (2010) A survey of online failure prediction methods. ACM Comput Surv 42(3):1–42

Snodgrass R, Yao S, Collberg C (2004) Tamper detection in audit logs. In: Proceedings of the 30th international conference on very large data bases, Toronto, 29 Aug–3 Sept 2004

Thimm H (2015) IT-supported assurance of environmental law compliance in small and medium sized enterprises. Int J Comput Inf Technol 4(2):297–305

Walker B, Redmond J, Sheridan L, Wang C, Goeft U (2008) Small and medium enterprises and the environment: barriers, drivers, innovation and best practice—a review of the literature. Small and Medium Enterprise Research Centre, Edith Cowan University, Western Australia

Designing for Engagement: A Case Study of an ICT Solution for Citizen Complaints Management in Rural South Africa

Carl Jacobs, Ulrike Rivett, and Musa Chemisto

1 Introduction

This study forms part of South Africa's Water Research Commission (WRC) project K5/2114 that investigated the possibility of incentivizing community engagement in order to improve drinking water supplies in South Africa. The research was based on the hypothesis that increasing community engagement will lead to an increased understanding of the current challenges of supplying drinking water to under-resourced communities. It was also suggested that through improved communication between communities, Water Service Authorities (WSAs) and Water Service Providers (WSPs) the 'experience of greater transparency and accountability for all stakeholders' would increase (Rivett et al. 2013).

In order to facilitate and improve the communication between stakeholders, ICTs (Information and Communication Technologies) were identified as a potential solution. It had been the original intention of the project to develop a mobile application that could be used by community members to log complaints remotely from their handset. However, the study showed that this did not align with the needs of the communities and municipalities. This paper reflects on the process of co-design in developing a complaints mechanism in rural environments.

C. Jacobs (✉) • U. Rivett • M. Chemisto
University of Cape Town, Cape Town, South Africa
e-mail: carljacobsza@icloud.com; ulrike.rivett@uct.ac.za; msxche002@myuct.ac.za

J. Marx Gómez, B. Scholtz (eds.), *Information Technology in Environmental Engineering*, Springer Proceedings in Business and Economics,
DOI 10.1007/978-3-319-25153-0_6

2 Research Objectives and Methodology

The objectives of the study were to:

1. Establish the status quo of current engagement practices between municipalities and communities and to identify the challenges facing municipalities and communities when complaints relating to water and sanitation provision are made.
2. Develop and implement a sustainable and context appropriate ICT system that improves complaints management and the engagement between communities and their municipalities.

Two municipalities with seven communities in rural Eastern Cape of South Africa were identified as the study sites. The criteria for the site selection were based on the rurality index of South Africa (Rivett et al. 2013) highlighting aspects such as population density, illiteracy levels, unemployment, social grant dependency etc.

A co-design approach was used throughout the study. Co-design emphasizes engagement between those responsible for delivering a service, stakeholders and users (Cruickshank and Deakin 2011). Since the community and municipalities were co-designers, the needs of each stakeholder were represented in the definition of the design and paternalistic notions of development were avoided by engaging with the communities and municipalities directly (Rivett et al. 2014b). This resulted in important partnerships, between the research team, the municipalities and the respective communities, being built.

Semi-structured interviews with municipal staff members were held between March 30th 2014 and April 4th 2014. Five topics were covered: municipal structure, water and sanitation services, service delivery, customer relations, and existing information systems. Similarly, semi-structured interviews and focus group engagements covering the same five topics were held with community members between May 26th 2014 and May 30th 2014.

The data analysis workshops and brainstorm sessions, both within the research team and with expert stakeholders, resulted in a first conceptualization of the system. The conceptual solutions were presented to each municipality between the 14th and 16th July 2014. After a second iteration, the project and its findings were presented to municipal stakeholders and members of the research community on 17th July 2014.

The system was introduced to the municipalities and their communities between October 27th and October 30th 2014. Communities were informed how the system worked and what avenues for laying complaints existed. Workshop sessions with municipal staff members were held. Each municipality and its communities were asked to begin using the system on Monday the 3rd November 2014. At the time of writing, the ICT system had only been in use for 2 months. The data presented in this paper reflects only on this time period.

3 Literature Review

In South Africa, public participation is a constitutional right. Tsatsire (2008) argues that public participation must be pursued, not only to comply with legislative prescriptions, but also to promote good corporate governance.

International experience has shown that citizen and community participation is an essential part of effective and accountable local governance (Shaidi 2007). Although part of the new developmental mandate assigned to local government, public consultation and participation remain a challenges confronting municipalities. Based on previous research, a hindrance to reporting water supply faults by community members has been the limited understanding of roles and responsibilities of local and district municipalities (Rivett et al. 2013).

Hohmann and Janssen (2012) showed that individuals are more committed to co-operate if communication between stakeholders increases and if social feedback reinforces the co-operative nature of an individual. Literature studies show that ICTs provide a means for improved governance and citizen engagement by providing access to previously unobtainable information (Asgarkhani 2005; Bertot et al. 2010).

However, sustainability of ICT applications is low and there is a high degree of failure of projects. The success or failure of an ICT project can depend on the design-reality gap that exists between 'current realities' and 'design conceptions of the e-government system' (Heeks 2002a) as well as the institution's adaptive capacity (Gupta et al. 2010). Heeks (2002b) also highlights that ICT systems that cannot address prioritized community needs tend to fail. ICT systems that are context relevant and characterized by simplicity are likely to minimize the design-reality gap and remain sustainable to the institutions and the people for which they are developed (Rivett et al. 2014a).

4 Analysing Existing Engagement Practices

Municipalities and communities have established practices in which they engage with each other. In order to develop a system, the study had to understand existing practices to enhance the enablers to communication and minimize the barriers. Unsurprisingly, the challenges identified in the local municipalities were very similar and can be categorized as follows:

Method of communication:
The most common methods used by citizens to engage with their municipality were phone calls and "walk-ins", i.e. walking to the municipal office. However, citizens often did not know who to complain to and/or laid complaints with the wrong department. They were also deterred by the financial costs associated with laying a complaint.

Municipalities communicated with citizens through community meetings, via their public communications officers or by posting announcements. Loud hailers were also often used to inform communities of upcoming meetings.

Method of lodging and tracking complaints:
Complaints were recorded either on a simple excel spreadsheet or in a logbook. No reference numbers or other tracking mechanisms were used to establish the status of a complaint. Both municipalities and their communities confirmed the lack of feedback. The perception of preferential treatment was also highlighted by communities.

Method of responding to complaints:
Both municipalities were very resource constrained and tasks were allocated on a priority basis. Aging infrastructure and the large distances between towns added to the challenge of resolving matters quickly. The internal workflow processes were informal and relied on institutional knowledge and relationships between staff.

In addition to establishing the challenges, enablers to implementing an ICT system were identified. Both municipalities had existing IT infrastructures in place. Technical teams, dedicated to resolving issues of water and sanitation, were available. Both municipalities displayed a strong willingness to be a part of the research project and there was a definite interest in taking ownership of the system. Based on the assessment of the current engagement practices, a number of important design features were identified:

- The reporting of complaints and issues had to be possible at any time, day or night.
- The reporting had to be free for citizens.
- The system had to allow for reporting at different locations, including satellite offices.
- There was a clear need for a feedback loop to include all relevant parties that respond to a complaint.

5 System Design and Development

Based on the findings highlighted above, a conceptual design of the system was developed. This is presented, along with descriptions, in (Fig. 1).

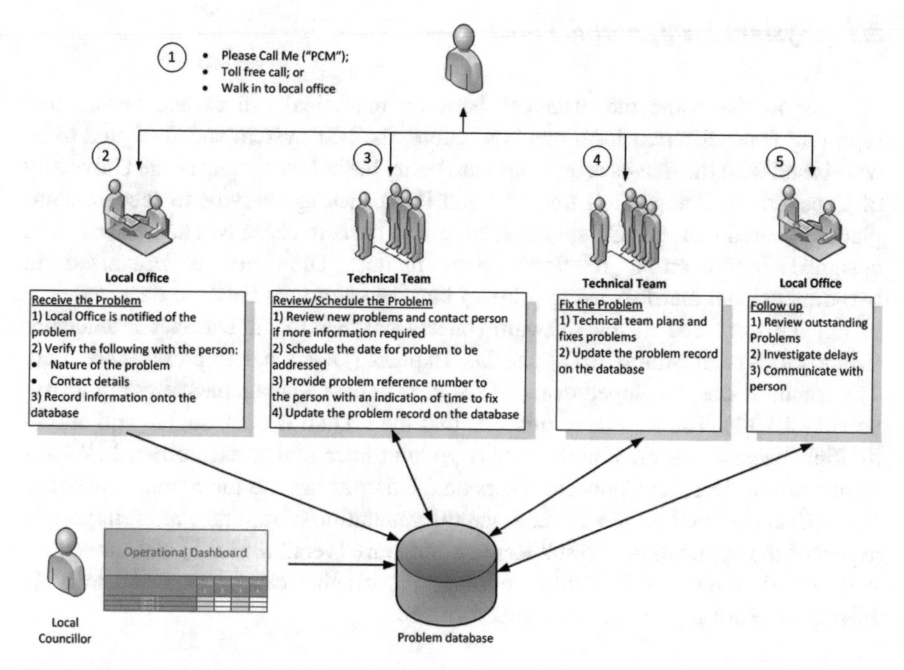

Fig. 1 Showing the conceptual system design

1. Citizens are able to use either a Please-Call-Me (PCM), a toll free line or can walk to the office to register their complaint.
2. The local office (which can also be a satellite office) will receive the problem and register the nature of the problem as well as the contact details. This information will be recorded on the database.
3. The technical team will log into the database daily to review new problems. The team will schedule a date for the problem to be addressed by and the system will create a reference number that will be sent via SMS to the citizen.
4. The technical team will update the record on the database once the details of the problem are clarified.
5. If the problem is still outstanding within 3 days, the local office will review the problem, investigate the delays and communicate with the citizen.
6. Local councillors or community development workers can log into the database to review records.

While commercial systems that could be used for this context are available, e.g. automated call center applications, software systems and off-the-shelf mobile applications, detailed investigation showed that the systems were unsuitable for three main reasons: they required substantial customization, a high level of technical skill and were expensive. Both municipalities had indicated that there was no budget for any IT software and that it was not foreseeable that this situation would change. It was therefore decided with the users that a low cost and low-maintenance system would be the most appropriate.

5.1 System Design and Tools

In order to overcome the distances between municipal offices and ensure that reporting from different locations is possible, the ICT system was designed to be web-based with the database and software being hosted on servers at the University of Cape Town. Using a toll free line and PCM message service to receive complaints ensured that the ICT system incurred no costs to citizens. The database was designed based on a relational data model. The data is organized in two-dimensional matrices using primary keys as identifiers between datasets.

For the front end of the software (interfaces), Microsoft Dot Net Framework (Microsoft Visual Studio 2012) and DevExpress (Developer Express) were used. The database was developed using MYSQL Database Management System (client version 5.1.11). The benefit of this was that the system is open source and allows multiple users to log on simultaneously without interrupting each other. MYSQL-Visual Studio Dot Net Connector (version 6.6.5) was used to interconnect between MYSQL and Visual Studio 2012. Since the generation of reports was an important aspect of the application, Crystal Reports Software (version 13.0.5) was embedded with visual studio and is used to produce reports that can be exported to PDF, Microsoft Word and Microsoft Excel formats.

5.2 System Accessibility

Users at the municipality access the system through the web URL address *amanzi.uct.ac.za*. System usage is restricted to municipal staff (User) and the iCOMMS research team (Admin). Figure 2 indicates the roles available to each user case scenario.

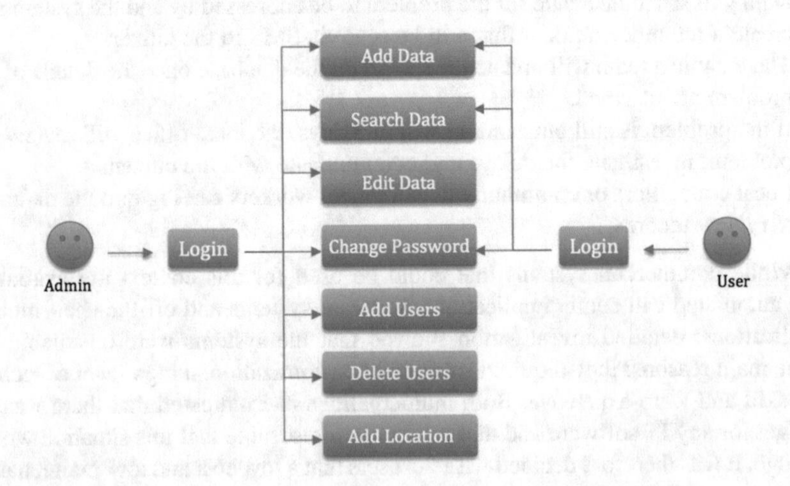

Fig. 2 System access

6 Findings and Discussion

At the time of writing this paper, the online complaints database system had been running for just over 2 months. The data and observations pertaining to November and December 2014 are presented (Fig. 3).

The data shows that community members started using the system immediately through the various channels. Whilst previous records were limited and make a numerical comparison difficult, the number of complaints suggests that there was an increase. Community members were clearly comfortable using the system to complain and this was confirmed in subsequent interviews held in March 2015.

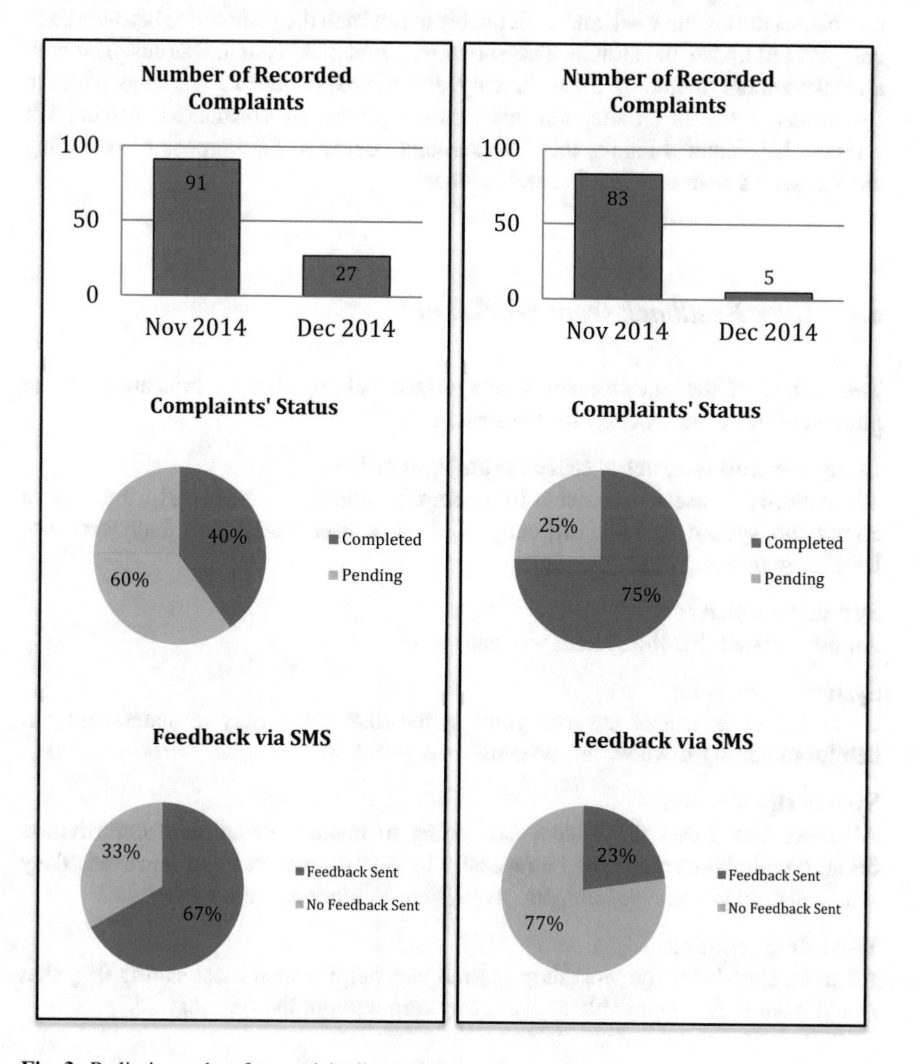

Fig. 3 Preliminary data for municipality 1 (*left*) and 2 (*right*)

The decline in recorded complaints during December was likely due to municipal staff taking leave over the festive season leading to fewer complaints received during this period. It is unclear if *pending* complaints were still being attended to or if their statuses had not been updated on the system. For the complaints where no feedback was sent, either no contact number was entered or the number was invalid i.e.: entered incorrectly. The reason for no contact number being entered was not clear. Citizens laying complaints may not have owned or had access to a phone or the user of the system may have forgotten to ask for or record a contact number. The lack of feedback experienced in Municipality 2 was due to the main office developing a "workaround" to the system. Complaints were uploaded to the online system in batches every Monday morning. This was because technical teams were out in the field seeing to complaints during the week and were unable to return to the main office between each complaint to update the system. This shows that whilst the system was designed with all stakeholders (including the technical team) the experience of workings with the system can result in an adaption that was not previously considered. Although it assisted the technical teams, this "workaround" defeated the purpose of providing citizens with a reference number and feedback.

6.1 User Feedback from Workshops

The results of the questionnaires categorized below, give an indication of the municipalities experiences with the system:

Computer and internet access at municipal offices:
The majority of users had access to a computer and the internet and were able to access the system without difficulty. Only one user stated that they were not familiar with computers.

System simplicity:
All users stated that the system was easy to use.

System user guide:
Users that made use of the user guide stated that it was easy to understand and helpful in getting to know the system.

System significance
All users stated that the system was going to make a meaningful contribution because complaints data could now easily be stored, managed and retrieved. They also stated they would recommend the system to other municipalities.

Workshop training:
All users stated that the workshop session was helpful with most stating they that would have in fact been able to use the system without the training.

Areas of concern that were raised by both municipalities were slow/intermittent Internet connections and possible abuse of the PCM message service by staff members and citizens.

7 Conclusions

This paper presented the findings of using a co-design approach to developing a citizen complaints system for rural municipalities in South Africa. The original intention of the researchers and funder had been to develop a mobile application that would allow community members to lodge complaints remotely from their handset.

By making communities, municipality workers and experts in the water and sanitation field designers of the system, it became rapidly apparent that rather than a high-tech solution, the mere introduction of a system that allows a "free-at-the-point-of-use" solution was the preferred option.

The system that was subsequently developed has shown to have a high uptake by community members and the municipalities. By allowing the co-designers to have an equal voice in the system design, the agenda was less driven by technology and more focused on developing a solution that spoke to the needs and abilities of municipalities and communities.

Although using the co-design approach involved all stakeholders it came with its own challenges. The logistics of organizing meetings with all stakeholders present proved challenging. Decisions often had to be made without everyone available and this, understandably, left those absent feeling excluded and/or their opinions disregarded.

Both municipalities indicated an interest in expanding the system to include not only complaints relating to water and sanitation but also other municipality services. Given that the system's design is based on simplicity and ease of use, transferring it to other application domains would not be difficult.

It is imperative to note that co-design, as an approach for software solutions, is more time and resource intensive at the outset of a project. Ongoing engagement requires commitment, time and finances (particularly in remote areas). It also requires flexibility of the researchers, developers and funders to accept that simple solutions might be a sustainable alternative.

Acknowledgments We would like to thank the Water Research Commission (WRC) for supporting this research project. We would like to express our sincere gratitude to the municipalities and community members in the Eastern Cape for their support of this research.

References

Asgarkhani M (2005) The effectiveness of e-service in local government: a case study. Electron J E-Gov 3(4):157–166

Bertot J, Jaeger P, Grimes J (2010) Using ICTs to create a culture of transparency: E-government and social media as openness and anti-corruption tools for societies. Gov Inf Q 27(3):264–271

Cruickshank P, Deakin M (2011) Co-design in smart cities. http://www.smartcities.info/files/Co-Design-in-Smart-Cities.pdf. Accessed 12 June 2015

Gupta J, Termeer C, Klostermann J, Meijerink S, van den Brink M, Jong P, Nooteboom S, Bergsma E (2010) The adaptive capacity wheel: a method to assess the inherent characteristics of institutions to enable the adaptive capacity of society. Environ Sci Pol 13(6):459–471

Heeks R (2002a) eGovernment in Africa: promise and practice. http://www.seed.manchester.ac.uk/medialibrary/IDPM/working_papers/igov/igov_wp13.pdf. Accessed 12 June 2015

Heeks R (2002b) Information systems and developing countries: failure, success, and local improvisations. Inf Soc 18(2):101–112

Hohmann N, Janssen M (2012) Addressing global sustainability challenges from the bottom up: the role of information feedback. The center for the study of institutional diversity resides in the School of Human Evolution and Social Change at Arizona State University

Rivett U, Taylor D, van Belle J-P, Chigona W, Maphazi N, Forlee B (2013) An assessment of incentivising community engagement in drinking water supply management. http://www.wrc.org.za/Pages/DisplayItem.aspx?ItemID=10676&FromURL=%2fPages%2fKH_AdvancedSearch.aspx%3fdt%3d%26ms%3d%26d%3dCommunity+engagement+in+drinking+water+supply+management%3a+a+review+%26start%3d1. Accessed 12 June 2015

Rivett U, Marsden G, Blake E (2014a) ICT for development: extending computing design concepts. In: Cooper B, Morrel R (eds) Africa-centred knowledges-crossing fields and worlds. Boydell & Brewer Ltd, Suffolk, pp 126–141

Rivett U, Taylor D, Chair C, Forlee B, Mrwebi M, van Belle J-P, Chigona W (2014b) Community engagement in drinking water supply management: A Review. WRC Report No. TT 583/13, ISBN 978-1-4312-0506-8. www.wrc.org.za

Shaidi EW (2007) An investigation of the role of Motherwell Ward Committees in influencing community participation for the period 2000 to 2006: Motherwell, Nelson Mandela Bay. Dissertation, Nelson Mandela Metropolitan University

Tsatsire I (2008) A critical analysis of challenges facing developmental local government: a case study of the Nelson Mandela Metropolitan Municipality. Dissertation, Nelson Mandela Metropolitan University

An Analysis of the Perceived Benefits and Drawbacks of Cloud ERP Systems: A South African Study

Brenda Scholtz and Denis Atukwase

1 Introduction

To stay competitive businesses have to strategically manage their resources which can be done by the adoption of Enterprise Resource Planning (ERP) systems. ERP systems can be used as a strategic resource by companies to gain competitiveness through integration of business processes and optimisation of available resources (Maditinos et al. 2012). ERP systems work essentially at integrating inventory data with financial, sales, and human resources data, allowing organisations to price their products, produce financial statements, and manage effectively their resources of people, materials, and money (Turban et al. 2010). An ERP system can be hosted on site or in an offsite server in the cloud. For an ERP system hosted in the cloud, the host servers for the ERP system are not physically at the premises of the end user, and if the end user just pays for the right to use the software, then it is called software as a service (SaaS), which is a form of cloud computing (Berman et al. 2012; Mell and Grance 2011). Due to cost pressures on companies, there is a growing trend of adoption of cloud computing systems especially SaaS.

A cloud ERP system allows an organisation to benefit from the ERP system without having to purchase and maintain the entire Information Technology (IT) infrastructure (Grubisic 2014). The implementation of a cloud ERP solution rather than a traditional on-site ERP system provides benefits such as reduced overhead costs, a significant reduction in time to implement the system and improved cash flow for the business via a subscription mechanism for use of the ERP system as the cost is operational rather than a large upfront capital expenditure.

B. Scholtz (✉) • D. Atukwase
Department of Computing Science, Nelson Mandela Metropolitan University (NMMU), Port Elizabeth, South Africa
e-mail: brenda.scholtz@nmmu.ac.za

J. Marx Gómez, B. Scholtz (eds.), *Information Technology in Environmental Engineering*, Springer Proceedings in Business and Economics,
DOI 10.1007/978-3-319-25153-0_7

ERP can be a source of competitive advantage, attracting new customers and revenue streams (Berman et al. 2012; Ragowsky and Gefen 2008).

A research study performed by IDC (2013b) shows that the cloud IT market is growing at five times the growth of the overall IT industry. The biggest usage and growth of this market is in developed economies, growing at 43.6 % (2011–2016). IDC (2013b) predicts that emerging economies will represent about 25 % of this market by 2016 with Africa and the Middle East representing 1 % of this market. South Africa is an emerging economy, so there is a predicted growth potential for this market. In a study done in Europe (France, United Kingdom, Germany, Spain and Poland) by Avrane-Chopard and Meunier (2011) it was reported that companies requiring IT services were considering using cloud solutions for various IT needs. The number of companies considering using a cloud solution was greater than those who were against it. Although the study was of European companies, it further illustrates the premise that cloud computing services are seriously being considered by companies as alternatives to traditional IT solutions. In the same study it was found that nearly 80 % of companies that adopt cloud solutions go on to adopt more cloud services or systems. This could prove the argument that there are benefits to cloud solutions and that the benefits outweigh the disadvantages (Avrane-Chopard and Meunier 2011). However with the trend towards sustainability several studies are investigating the environmental impact of cloud computing. Cloud computing can address two areas of a green IT approach: energy efficiency and resource efficiency by using technologies such as resource virtualisation and workload consolidation (Priya et al. 2013). However the data centres they utilise require high energy usage for its operation.

The increase in cloud IT solutions has led to the growth of the cloud ERP market. Braund (2014) predicted that the growth of the cloud ERP market in 2013 will remain consistent in 2014. Braund (2014) also noted that there is an increase in demand for cloud ERP expertise, supporting the notion that cloud ERP usage is growing. On the other hand a study by Panorama Consulting Solutions (Panorama Consulting 2014) reported a drop in cloud ERP adoption. This study also reported that the main reasons for non-adoption of cloud computing are lack of knowledge about cloud offerings and fear of the risk of security breaches. In a study of cloud adoption by small and medium-sized enterprises (SMEs) the main reason for not adopting cloud ERP systems were security concerns (Castellina 2012a).

In a study by World Wide Worx (2014), adoption of cloud computing by SMEs in South Africa had grown from between 9 % in 2012 to 27 % in 2014. So there is growth of cloud computing adoption in South Africa. However, in a study about the awareness of cloud computing by SMEs in South Africa (Mohlameane and Ruxwana 2014), the understanding of cloud computing by participants in the study ranged from basic to low understanding (80 % of respondents). In the same study respondents identified the top challenges of cloud computing to be the performance and availability of cloud services because they have to be accessed over the Internet.

There is a gap in research related to cloud ERP systems and particularly in developing countries like South Africa which have additional technology issues like bandwidth. Although a few studies have been done related to the adoption of cloud computing in South Africa, the perceived benefits and drawbacks of cloud ERP systems in South Africa is not known.

The aim of this paper is to explore the cloud ERP systems literature and analyse the benefits and drawbacks of cloud ERP systems. A theoretical model is derived and empirically validated by means of a survey of South African organisations. The structure of this paper is as follows: Sect. 2 discusses the importance and growth of both the ERP system market as well as the cloud computing market. In Sect. 3 the benefits and drawbacks of cloud ERP systems as identified in other studies is examined and a theoretical model derived. The methodology and objectives of the research is discussed in Sect. 4. Section 5 provides an analysis of both the qualitative and quantitative results whilst Sect. 6 concludes with some recommendations.

2 Growth and Dominance of Cloud ERP Systems

An ERP system is an integrated suite of modules that forms the basis for operational and transactional records for an enterprise (Castellina 2012b). In the early years of ERP systems and its predecessors, it was mainly adopted by large-sized enterprises (LEs) especially in the manufacturing sector. ERP systems are now widely adopted in a range of industries to manage business processes. The cost of hosting, implementing and using these systems was originally prohibitive for SMEs. However, SMEs are driving the change in new IT concepts and this has led to an increase in the use of ERP systems in SMEs (Grubisic 2014). According to Gartner (Columbus 2014) the worldwide ERP software market grew 3.8 % from $24.4 billion in 2012 to $25.4 billion in 2013. The biggest driver of the various data reported was growth of ERP use by SMEs (Hoseini 2012; Jacobson et al. 2007) and companies with less than 5 years in business (IDC 2013b). ERP spending worldwide is projected to grow from $26.03 billion in 2013 to $34.3 billion in 2017, attaining a compound annual growth rate (CAGR) in the forecast period 2012–2017 of 7 % (Columbus 2014). ERP systems can lead to greater efficiency in doing business and a more coherent flow of information with all departments in an enterprise using the same system. It eliminates duplication of information (Maditinos et al. 2012; O'Leary 2004). However in spite of the potential benefits of ERP systems a survey by Panorama Consulting Solutions (2014) revealed that 66 % of respondents indicated that measurable benefits on their ERP solutions were less than 50 % of what they had anticipated. A general trend of exceeding ERP project budget costs and duration overruns is also reported. This can be a significant problem for companies with limited resources. Therefore, companies should carefully plan their ERP implementation projects to avoid these problems. The project and cost overruns contribute to realisation of fewer benefits than expected. The Panorama study also reported that the number of respondents using SaaS and cloud ERP systems decreased from 2013 to 2014 (from 26 % to 15 %).

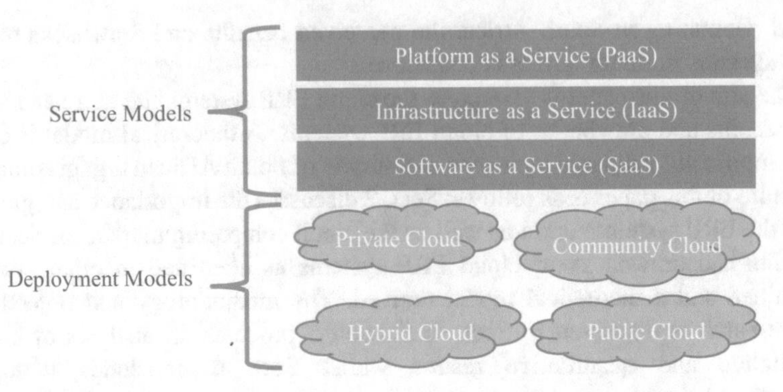

Fig. 1 Cloud service and deployment models (Mell and Grance 2011)

Bughin and Manyika (2012) define cloud computing as the ability to access highly scalable computing resources through the Internet, often at lower prices than those required to install on one's own computers because the resources are shared across many users. There are several other definitions of cloud computing, but the most standardised version is by the US Department of Commerce National Institute of Standards and Technology (NIST). NIST defines cloud computing as a model for enabling ubiquitous, convenient, on-demand network access to a shared pool of configurable computing resources (for example, networks, servers, applications and services) that can be rapidly provisioned and released with minimal management effort or service provider interaction (Mell and Grance 2011). There are three types of cloud computing (Mell and Grance 2011). The three types are (Fig. 1), Platform as a Service (PaaS); Infrastructure as a Service (IaaS); and Software as a Service (SaaS). Platform as a Service (PaaS) is the capability to deploy onto the cloud infrastructure consumer-created or acquired applications created using programming languages, libraries, services, and tools supported by the provider. The consumer does not manage or control the underlying cloud infrastructure including network, servers, operating systems, or storage, but has control over the deployed applications and possibly configuration settings for the application-hosting environment. Infrastructure as a Service (IaaS) is the provision of processing, storage, networks, and other fundamental computing resources where the consumer is able to deploy and run arbitrary software, which can include operating systems and applications. The consumer does not manage or control the underlying cloud infrastructure but has control over operating systems, storage, and deployed applications; and possibly limited control of select networking components (for example, host firewalls). Software as a Service (SaaS) is the capability provided to the consumer use the provider's applications running on a cloud infrastructure. The applications are accessible from various client devices through either a thin client interface, such as a web browser or a program interface. The consumer does not manage or control the underlying cloud infrastructure including network, servers, operating systems, storage, or even individual application capabilities, with the possible exception of limited user-specific application configuration settings.

Cloud computing is further subdivided into the following categories: private cloud; community cloud; public cloud; and hybrid cloud. A private cloud is for limited use by a single organisation with multiple users. It may be owned, managed, and operated by the organisation, a third party, or some combination of them, and it may exist on or off premises. A community cloud is for limited use by a specific community of users from organisations that have shared concerns (for example, mission, security requirements, policy, and compliance considerations). It may be owned, managed, and operated by one or more of the organisations in the community, a third party, or some combination of them, and it may exist on or off premises. A public cloud is for open use by the general public. It may be owned, managed, and operated by a business, academic, or government organisation, or some combination of them. It exists on the premises of the cloud provider. A hybrid cloud is a composition of two or more distinct cloud infrastructures (private, community, or public) that remain unique entities, but are bound together by standardised or proprietary technology that enables data and application portability (for example cloud bursting for load balancing between clouds).

By 2015, cloud computing could represent a $70–85 billion opportunity, with the market set to double every 2 years. In estimation, the SaaS market was predicted to reach $63 billion in 2014 (Eval-Source 2014). Some technology analysts forecast that by 2015 cloud computing infrastructure and applications could account for 20 % of total spend in these areas. IDC (2013a) predicts that by 2017 the size of the cloud computing market will be about $107 billion, up from $47 billion in 2003 with SaaS representing the biggest share of this market. This is a compound growth of 23.5 %, which is five times that of the IT industry (IDC 2013a).

In a study done by the Aberdeen Group, between 2009 and 2013, there was a steady increase in businesses which were considering using ERP SaaS at their next ERP implementation and a decrease in the traditional on site ERP solutions (Castellina 2014a). However the majority of implementations in the study still use traditional licensed on-site ERP. Whilst SaaS ERP has increased from 23 % in 2009 to 47 % in 2013, traditional on-site ERP decreased from 80 % to 60 % in the same period. In the same Aberdeen Group study, it was found that with newer ERP installations amongst wholesalers and distributors (less than 2 years old), SaaS was more popular and with older installations (greater than 2 years old), traditional on-site licensed ERP systems were more popular.

Adoption of ERP SaaS has generally been slow compared to other cloud applications such as Customer Relationship Management (CRM) and Human Resource Management (HRM) (Clear Water Corporate Finance 2013). As businesses familiarise themselves with cloud applications in categories such as CRM and HRM, then SaaS ERP has better prospects of being adopted. This view is supported by Lechesa et al. (2011) who put forward that as organisations adopt peripheral services via cloud computing, it will eventually lead to procuring all services in the cloud. Even with increased use of SaaS, there are still security concerns and negative perceptions or barriers to adoption that limit its uptake by companies. While, security concerns is the top reason for not adopting SaaS, in reality research has shown that on site ERP has more security issues as compared to cloud SaaS ERP (Castellina 2012a; Utzig et al. 2013). Cloud ERP service providers have heavily invested in ensuring that their data centres are well secured against any perceived threat. A security breach can be a public relations nightmare which is one of the reasons why they are investing millions of dollars on security. However, more needs to be done by cloud services providers to allay these consumer concerns as they are just perceptions. In addition, consumers of cloud services should carefully select their service providers and choose those that are audited by a third party, which should go some way in allaying the security fears.

3 Benefits and Drawbacks of Cloud ERP Systems

Businesses stand to gain a lot from cloud computing services. They can get access to infrastructure and software that they pay for on an on-going basis which they would otherwise not be able to afford. Therefore, cloud computing offers businesses the opportunity to improve their IT capabilities in a way that they previously could not (Diamadi et al. 2011). Cloud computing separates IT resources, such as files and programs, from the devices used to access them which can create many advantages, such as resource pooling and ability to rapidly upscale or reduce available resources (Bughin and Manyika 2012). This gives businesses the flexibility to scale up or down on their IT and to make changes quicker. This is essential as the modern global market place requires companies to be easily flexible and adaptable to change. This is more relevant to SMEs as they are more adaptable to take advantage of the changing environment as compared to LEs. They are the driving force encouraging the IT industry to bring new concepts and innovations such as cloud computing.

An analysis of studies citing the benefits of cloud ERP systems revealed nine benefits which were reported the most frequently (Table 1). The most often cited benefit is the reduced costs of implementation. Cloud services reduce IT costs since they eliminate the need for in-house IT infrastructure to support the ERP systems. A third party hosts the ERP system in the cloud on their behalf. This significantly reduces upfront and operating costs (Aleem and Sprott 2013; Avrane-Chopard and Meunier 2011; Castellina 2013, 2014b; Duan et al. 2013; Hoseini 2012; Utzig et al. 2013; Yang 2012). This is particularly useful for companies (such as SMEs) that rely on a cost effective ERP system. Due to the low cost of implementation and use of cloud ERP systems, they can be funded as an operational and not as a capital expense. Some organisations cannot afford elaborate security to their IT infrastructure. A breach of data can have disastrous consequences. On the other hand, cloud service providers can afford to invest in robust security for their IT, thereby benefiting their customers (Aleem and Sprott 2013; Avrane-Chopard and Meunier 2011; Duan et al. 2013; Rader 2012).

With cloud ERP systems, organisations can focus on their core business and let a third party manage the ERP system for them. This could lead to improved productivity within the business (Avrane-Chopard and Meunier 2011; Duan et al. 2013;

Table 1 Benefits of cloud ERP systems

Benefits	Reference literature
Reduced IT costs	Aleem and Sprott (2013), Avrane-Chopard and Meunier (2011), Castellina (2014b), Duan et al. (2013), Hoseini (2012), Utzig et al. (2013), Yang (2012)
Focus on the core business	Avrane-Chopard and Meunier (2011), Duan et al. (2013), Yang (2012)
Access to latest ERP developments	Avrane-Chopard and Meunier (2011), Castellina (2014b), Duan et al. (2013), Utzig et al. (2013), Yang (2012)
Improved IT security	Aleem and Sprott (2013), Avrane-Chopard and Meunier (2011), Duan et al. (2013), Rader (2012)
Flexibility	Aleem and Sprott (2013), Avrane-Chopard and Meunier (2011), Castellina (2014b), Duan et al. (2013), Hoseini (2012), Rader (2012), Utzig et al. (2013)
Improved business efficiency	Castellina (2013, 2014b)
Easier scalability	Castellina (2014b), Rader (2012), Utzig et al. (2013)
Decreased data execution time	Aleem and Sprott (2013)
Improved collaboration	Castellina (2013, 2014b)

Yang 2012). As ERP systems are developed and improved by ERP vendors, organisations gain from these advancements without significant cost to them (Avrane-Chopard and Meunier 2011; Castellina 2014b; Duan et al. 2013; Utzig et al. 2013; Yang 2012). Cloud services give organisations flexibility because they can be accessed from anywhere that has an Internet connection. This increases productivity as one does not have to physically be in the office to access the ERP system. It also ensures business continuity as data is backed up on these systems and can be accessed if there is an emergency such as fire at the organisation's premises (Aleem and Sprott 2013; Avrane-Chopard and Meunier 2011; Castellina 2013, 2014b; Duan et al. 2013; Hoseini 2012; Rader 2012; Utzig et al. 2013). Due to cloud ERP systems, business efficiency is improved since the decision making process in the organisation is made quicker and easier due to ease of access to data. All data such as sales figures and inventory is stored in one system that can be accessed from anywhere. This data storage system also leads to improved collaboration between the different divisions of the business and their customers (Castellina 2013, 2014b). Improved collaboration of the divisions in different locations helps to better serve customers.

Cloud ERP systems lead to easier scalability of the ERP system if the business grows. There is no requirement from the end user to invest in new infrastructure. This is done by the cloud ERP vendor or host (Castellina 2013, 2014b; Rader 2012; Utzig et al. 2013). Cloud ERP systems have the ability to decrease data execution time since a lot of data can be processed in a shorter time (Aleem and Sprott 2013). This ability is one of the main reasons organisations opt towards the cloud. For all the benefits of cloud computing, several studies have mentioned the disadvantages (Table 2). These disadvantages could responsible for some organisations not

Table 2 Drawbacks of cloud ERP systems

Drawbacks	Reference literature
Additional costs	Duan et al. (2013), Utzig et al. (2013)
Security risks	Duan et al. (2013), Yang (2012)
Lack of clear service level agreement	Duan et al. (2013)
Limited customisation and integration options	Duan et al. (2013), Utzig et al. (2013)
Strategic risks (IT outsourcing)	Duan et al. (2013)
Loss of IT competencies due to outsourcing	Duan et al. (2013)
Increased downtime	Yang (2012)

adopting cloud ERP. One of the most frequently reported disadvantages of cloud ERP systems are the major additional costs during implementation (Duan et al. 2013; Utzig et al. 2013). Cloud ERP systems offer basic functionality and generally don't offer extensive customisation which on-site traditional ERP systems offer. Therefore, any elaborate customisation will come at a high cost. Another frequently reported disadvantage is the security risks which are outside the control of the end user (Duan et al. 2013; Yang 2012). This is particularly concerning for the end users because ERP systems are an integral part of their business processes. A security breach can have disastrous consequences such as loss of data. With cloud ERP systems there is a risk of not having a clear performance based service level agreement (SLA) on which to base subscription expenses (Duan et al. 2013).

Cloud ERP systems have strategic risks because part of the IT department is outside the control of the organisation (Duan et al. 2013). The end user has to trust that the service provider will run and maintain the system with integrity and ensure good reliability. With outsourcing, companies might be compelled to reduce the number of employees in their IT department and outsource instead. Slow Internet speeds and an Internet service outage can lead to ERP system downtime for the client (Yang 2012). Because cloud ERP systems are accessed via the Internet, the end user does not have control over the speed of the Internet connection. This can lead to loss of productivity and lead to a complete lack of access to the cloud ERP system. Before adoption of cloud ERP systems, it is imperative for companies to evaluate the advantages and disadvantages of cloud ERP systems along with the companies' internal resources and capabilities. This will guide the companies on the type of cloud service and deployment model to adopt and improve the chances of ERP implementation success.

4 Research Methodology

The problem that this study addresses is that there is a lack of knowledge related to the extent of understanding of cloud ERP systems advantages and disadvantages. It is not evident how much companies know about cloud ERP systems and why or why not companies adopt cloud ERP systems (the reasons for adoption and non-adoption). The primary objective of this paper is to establish the advantages (benefits) and disadvantages (drawbacks) of cloud ERP systems as perceived by South African companies. The main research question is therefore "*What are the advantages (benefits) and disadvantages (drawbacks) of cloud ERP systems as perceived by South African companies?*"

A literature review was undertaken to create the theoretical model and a survey research strategy was adopted to empirically validate the model and answer the research question. The scope of the survey was limited to companies in South Africa and respondents were IT managers, line managers and IT professionals. Data was collected via an on-line questionnaire which was emailed to 49 companies. Of these 41 complete responses were received. The response rate cannot be measured because it is not clear how many companies received the questionnaire. The closed-ended questions were measured on a 5-point Likert scale (where 1 = strongly disagree and 5 = strongly agree). Respondents were given a set of statements about the advantages (benefits) and drawbacks (disadvantages) of cloud ERP systems. These were derived from the literature review. The following statistical ranges (categories) were applied to the analysed data: negative [1–2.60), neutral [2.61–3.4) and positive [3.41–5).

5 Analysis of Results

In terms of their employment status, the majority (27 %) of respondents were in middle management. Collectively 68 % of the respondents are in management, 22 % are senior employees and 10 % are junior employees (Fig. 2). Each respondent represented one company. The companies in the survey are from a wide range of industries and were classified into nine types of industry. The top three industries repented in the survey were IT (24 %), manufacturing (22 %) and engineering services (22 %). Of the 41 companies that took part in the survey, 15 (37 %) have more than 500 employees and in total 23 (56 %) have more than 200 employees. Therefore, 56 % (n = 23) of the companies that completed the survey are LEs and 44 % (n = 18) are SMEs. By definition, depending on the sector, in South Africa SMEs have less than 200 full time employees (National Small Enterprise Act 102 1996).

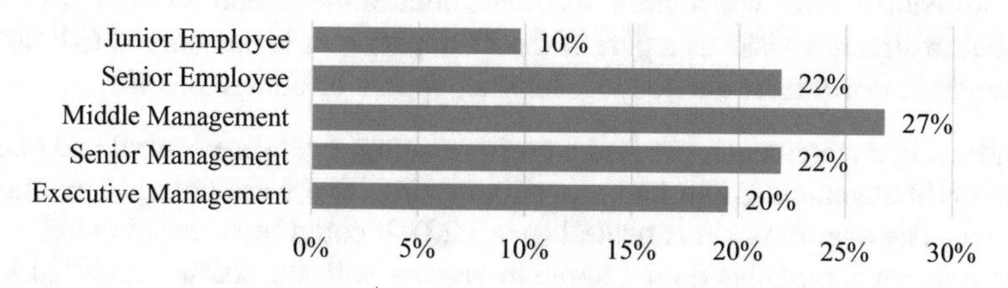

Fig. 2 Profile of respondents' status in the company (n = 41)

Table 3 Perceived benefits of cloud ERP systems (n = 41)

	Mean	S.D.	Strongly disagree/ disagree n	%	Neutral n	%	Agree/ strongly agree n	%
Flexibility	4.54	0.90	2	5	2	5	37	90
Improved collaboration	4.12	1.00	2	5	9	22	30	73
Improved business efficiency	4.02	0.91	1	2	10	24	30	73
Access to latest development in ERP systems	3.90	0.89	3	7	9	22	29	71
Scalability	3.90	0.97	4	10	9	22	28	68
Focus on core activities	3.83	0.89	2	5	14	34	25	61
Reduce IT costs	3.59	1.20	7	17	11	27	23	56
Decreased data execution time	3.34	0.76	5	12	19	46	17	41
Improved IT security	3.05	1.14	13	32	16	39	12	29

The biggest advantages of cloud ERP systems were identified from the survey and selected based on the percentage of respondents which Agree and Strongly Agree with the statements provided. The top three are (Table 3):

- Flexibility because they can be accessed from anywhere (90 %)
- Improved collaboration between the different divisions of the business and their customers (73 %)
- Improved business efficiency (73 %)

The overall mean is in the positive range ($\mu = 3.81$). Of the nine statements, seven had a positive rating and two were neutral. This shows that the respondents agree with the statements about the benefits of cloud ERP systems. Therefore it can be deduced that South African companies have a good understanding of the advantages of cloud ERP systems. These results agree with the study by Avrane-Chopard and Meunier (2011) as to why companies have adopted cloud computing. From the survey the top advantage of cloud ERP systems is to give the company flexibility and ease of access of the system.

Respondents were given statements about the disadvantages (drawbacks) of cloud ERP systems and they had to rate them on a 5-point Likert scale. The top three drawbacks of cloud ERP systems based on the percentage of respondents which Agreed or Strongly Agreed with the statements provided, are (Table 4):

- Pose security risks which are outside the control of the end user (51 %)
- Lead to strategic risks as a part of the IT department is outsourced (51 %)
- Increased down time due to slow Internet speeds or outage (49 %)

Of the eight statements, two had a positive rating, five were neutral and one was of the eight statements, two had a positive rating, five were neutral and one was negative. The overall mean is neutral ($\mu = 3.22$). It could be deduced therefore that South African companies do not agree in general with the studies analysed in this

Table 4 Perceived drawbacks of cloud ERP systems (n = 41)

			Strongly disagree/ disagree		Neutral		Agree/ strongly agree	
	Mean	S.D.	n	%	n	%	n	%
Security risks	3.61	1.159	9	22	11	27	21	51
Strategic risks	3.37	1.220	11	27	9	22	21	51
Increased down time	3.44	1.266	10	24	11	27	20	49
Loss of IT competencies	3.15	0.963	12	29	12	29	17	41
Additional implementation costs	3.20	1.209	12	29	14	34	15	37
Limited customisation options	2.83	1.138	17	41	12	29	12	29
Risk of not having a clear performance based SLA	2.93	1.127	12	29	17	41	12	29

paper. This could be due to differences in other countries or due to lack of real knowledge and understanding regarding the drawbacks of cloud ERP systems by South African small and medium sized companies. Future research is required in order to analyse this further.

6 Conclusions and Recommendations

The respondents have a good understanding of the benefits of cloud ERP systems. The top benefit of cloud ERP systems as perceived by South African companies is that it gives companies flexibility because they can be accessed from anywhere. The respondents' understanding of the disadvantages of cloud ERP systems is neutral. The top disadvantage of cloud ERP systems as perceived by South African companies is that it poses security risks which are outside the control of the end user. The paper provides a valuable contribution in terms of providing a more in depth understanding of the literature regarding cloud ERP systems, especially in developing countries such as South Africa. The benefits and drawbacks identified can be used to assist ERP vendors with marketing their cloud solutions. The results of the survey highlighted the perceptions of SMEs and larger companies in South Africa with regards to cloud computing. Future research could expand the study to larger companies.

References

Aleem A, Sprott CR (2013) Let me in the cloud: analysis of the benefit and risk assessment of cloud platform. J Financ Crime 20(1):6–24

Avrane-Chopard J, Meunier C (2011) Outlook—overcast and bright: how the cloud is transforming IT for SMBs. http://www.mckinsey.com/~/media/mckinsey/dotcom/client_service/telecoms/pdfs/recall_no18_01_outlook_cloud_computing.ashx. Accessed 1 Apr 2014

Berman SJ, Kesterson-Townes L, Marshall A, Srivathsa R (2012) How cloud computing enables process and business model innovation. Strateg Leadersh 40(4):27–35

Braund M (2014) ERP software in 2014: six predicted trends. http://www.huffingtonpost.co.uk/mark-braund/erp-software-in-2014-6-pr_b_4642399.html. Accessed 24 May 2014

Bughin J, Manyika J (2012) Internet matters: essays in digital transformation. http://www.mckinsey.com/insights/business_technology/essays_in_digital_transformation. Accessed 1 Apr 2014

Castellina N (2012a) SaaS and cloud ERP observations: is cloud ERP right for you? http://v1.aberdeen.com/launch/report/benchmark/7857-RA-enterprise-resource-planning.asp?lan=US. Accessed 24 Mar 2014

Castellina N (2012b) To ERP or not to ERP for SMBs: what can ERP do for me? http://v1.aberdeen.com/launch/report/benchmark/7798-RA-enterprise-resource-planning.asp. Accessed 24 Sept 2014

Castellina N (2013) SaaS and cloud ERP observations: enabling collaboration in the midmarket. http://v1.aberdeen.com/launch/report/benchmark/8754-RA-enterprise-resource-planning.asp?lan=US. Accessed 24 Mar 2014

Castellina N (2014a) SaaS ERP in wholesale and distribution: enabling communication across a wide network. http://v1.aberdeen.com/launch/report/benchmark/8832-RA-erp-wholesale-distribution.asp?lan=US. Accessed 24 Mar 2014

Castellina N (2014b) The benefits of cloud ERP: it's about transforming your business. http://v1.aberdeen.com/launch/report/research_report/8889-RR-erp-saas-cloud.asp?lan=US. Accessed 25 Apr 2014

Clear Water Corporate Finance (2013) Enterprise resource planning report 2013. http://www.clearwatercf.com/documents/sectors/ERP_Report_FINAL.pdf. Accessed 12 Feb 2014

Columbus L (2014) Gartner's ERP market share update shows the future of cloud ERP is now. http://www.forbes.com/sites/louiscolumbus/2014/05/12/gartners-erp-market-share-update-shows-the-future-of-cloud-erp-is-now/. Accessed 17 June 2014

Diamadi Z, Dubey A, Vora A (2011) Winning in the SMB cloud: charting a path to success. http://www.mckinsey.com/~/media/mckinsey/dotcom/client_service/highpercent20tech/pdfs/winning_in_the_smb_cloud.ashx. Accessed 1 Apr 2014

Duan J, Faker P, Fesak A, Stuart T (2013) Benefits and drawbacks of cloud-based versus traditional ERP systems. http://www.academia.edu/2777755/Benefits_and_Drawbacks_of_Cloud-Based_versus_Traditional_ERP_Systems#. Accessed 3 Apr 2014

Eval-Source (2014) ERP cloud and SaaS buyer's guide: version 3. http://www.eval-source.com/erp_cloud_and_saas_guide.html. Accessed 17 June 2014

Grubisic I (2014) ERP in clouds or still below. J Syst Inf Technol 16(1):62–76

Hoseini L (2012) Advantages and disadvantages of adopting ERP systems served as SaaS from the perspective of SaaS users. Dissertation, Pennsylvania State University

IDC (2013a) IDC forecasts worldwide public it cloud services spending to reach nearly $108 billion by 2017 as focus shifts from savings to innovation. http://www.idc.com/getdoc.jsp?containerId=prUS24298013. Accessed 17 June 2014

IDC (2013b) Successful cloud partners: HIGHER, FASTER, STRONGER. What IT solution providers need to know to build high-performing cloud businesses. http://www.microsoft.com/en-us/news/download/presskits/partnernetwork/docs/idcmicrosoftcloudinfodoc.pdf. Accessed 19 June 2014

Jacobson S, Shepherd J, D'Aquila M, Carter K (2007) ERP 2007 market sizing series: the ERP market sizing report, 2006–2011. http://sapfarm.com/news/AMR_ERP_Market_Sizing_2006-2011.pdf. Accessed 12 Feb 2014

Lechesa M, Seymour L, Schuler J (2011) ERP software as service (SaaS): factors affecting adoption in South Africa. In: Moller C, Chaudhry S (eds) Re-conceptualizing enterprise information systems. 5th IFIP WG 8.9 working conference, Aalborg, Denmark, Oct 2011. Lecture notes in business information processing, vol 105. Springer, Heidelberg, pp 152–167

Maditinos D, Chatzoudes D, Tsairidis C (2012) Factors affecting ERP system implementation effectiveness. J Enterp Inf Manag 25(1):60–78

Mell P, Grance T (2011) The NIST definition of cloud computing: recommendations of the National Institute of Standards and Technology. http://csrc.nist.gov/publications/nistpubs/800-145/SP800-145.pdf. Accessed 24 Mar 2014

Mohlameane M, Ruxwana N (2014) The awareness of cloud computing: a case study of South African SMEs. Int J Trade Econ Financ 5(1):6–11

National Small Enterprise Act 102 (1996) Government Gazette (17612). http://www.empowerdex.co.za/Portals/5/docs/act.pdf. Accessed 1 Mar 2014

O'Leary DE (2004) Enterprise resource planning (ERP) systems: an empirical analysis of benefits. J Emerg Technol Account 1:63–72

Panorama Consulting Solutions (2014) 2014 ERP report: a Panorama Consulting Solutions research report. http://go.panorama-consulting.com/2014-ERP-Report_Download.html. Accessed 11 Mar 2014

Priya B, Pilli ES, Joshi RC (2013) A survey on energy and power consumption models for greener cloud. In: Proceedings of the 2013 3rd IEEE international advance computing conference, Ghaziabad, India, 22–23 Feb 2013

Rader D (2012) How cloud computing maximizes growth opportunities for a firm challenging established rivals. Strateg Leadersh 40(3):36–43

Ragowsky A, Gefen D (2008) What makes the competitive contribution of ERP strategic. Data Base Adv Inf Syst 39(2):33–49

Turban E, Volonino L, McLean E, Wetherbe J (2010) Information technology for management: transforming organisations in the digital economy. Wiley, Asia

Utzig C, Holland D, Horvath M, Manohar M (2013) ERP in the cloud. Is it ready? Are you? http://www.strategyand.pwc.com/media/file/Strategyand_ERP-in-the-Cloud.pdf. Accessed 7 June 2014

World Wide Worx (2014) SME survey 2014. http://www.smesurvey.co.za/reports/SME%20Survey%20summary%202014 %204.pdf. Accessed 16 Nov 2014

Yang SQ (2012) Move into the cloud, shall we? Libr Hi Tech News 29(1):4–7

Collaborative Network Platform Solution for Monitoring, Optimization, and Reporting of Environmental and Energy Performance of Data Center

Gamal Kassem, Niko Zenker, Klaus Turowski, and Naoum Jamous

1 Introduction

The production of data center components entails significant environmental impacts. However, data center operators seldom consider these impacts in their purchasing decisions. Many other factors have to be considered in purchasing decisions. These include the different components life cycle phases that have on total environmental impact associated with the production, distribution and disposal of the component/ device. Aggregately considered as a result of energy efficiency gains in the use-phase of the new component, energy efficiency gains should be possessed by a new component, if the replacement of the older and less energy efficient component can be justified under the consideration of environmental concerns. For example, the results show that the production phase of some notebooks, which contains about 56 % (214 kg CO_2 emission in 5 years) of the total greenhouse gas emissions, creates a significantly higher impact than the use phase (Federal Environment Agency 2012). Moreover, the environmental impacts of the production phase of data center components are so high that they cannot be compensated in realistic time periods by energy efficiency gains in the use phase.

Awareness raised by societal and political entities is present in the form of memorandums and statements but lacks operable and deployable solutions. MORE is considered an operable solution which enables data centers to fulfill environmental policy and to consider societal awareness by measuring, reporting, and controlling EPIs such as CO_2 emissions, water waste, and the other environmental impacts of data center processes.

One of the relevant measurements which aims to control the power usage in data centers is Power Usage Efficiency (PUE). PUE was developed by The Green Grid

G. Kassem (✉) • N. Zenker • K. Turowski • N. Jamous
Otto-von-Guericke-Universität Magdeburg, Magdeburg, Germany
e-mail: kassem@ovgu.de; kontakt@zenker-it.de; klaus.turowski@ovgu.de; naoum.jamous@ovgu.de

J. Marx Gómez, B. Scholtz (eds.), *Information Technology in Environmental Engineering*, Springer Proceedings in Business and Economics,
DOI 10.1007/978-3-319-25153-0_8

association (TGG) in order to compute how much power is actually used by the computing equipment. In other words PUE is the ratio of the total amount of power used by a computer data center facility to its IT equipment power (The Green Grid 2008), PUE has become one of the most famous metrics in this field. In Belady et al. (2010) two other metrics are introduced which are intended to illustrate CUE for CO_2 emissions and WUE for water consumption.

Another tool to assess and evaluate the energy efficiency in data centers is the data center maturity model (DCMM). DCMM was invented in 2010 by The Green Grid (TGG) to provide clear capability descriptions of a data center so that owners/operators can benchmark their current performance; power usage, cooling requirements, computation ability, and storage capacity determine a data center's level of maturity. DCMM divides maturity into the levels from 0 to 5, with 0 being the minimum and 5 being the maximum, and it planned to be achieved in 2016 (The Green Grid 2011)

The IT industry, computer manufacturers, and governments are becoming increasingly concerned with the energy use of IT equipment. This is why the Standard Performance Evaluation Corporation (SPEC) developed SPECpower in 2007 a piece of benchmark software in order to enable manufacturers to better compete based on power consumption, and it also helps IT managers design data centers with more energy efficiency (SPEC 2012). SPECpower is the first step towards building energy efficient data centers because it concerns itself with the manufacturing of IT equipment at companies such as AMD, Dell, Intel, etc.

The European standard EN 50600 was invented in 2011 to address energy efficiency in data centers. It specifies the classification of systems based upon the key criteria of availability, security, and energy efficiency, and it specifies the measurement methodologies and report formats in monitoring the performance of data center facilities and infrastructures. The necessary management and operational information is specified in subsection EN 50600-2-X (Austrian Standards Institute 2012).

MORE aims to conduct all needed measurements on the aforementioned methods, frameworks, models, and standards related to energy consumption efficiency at the usage phase in data center sector. And it aims to also introduce them as standardized Usage Performance Indicators (UPIs).

Service Level Agreements (SLAs), IT regulatory compliance, and other requirements of IT service framework like the IT Infrastructure Library (ITIL) (Beims 2009) or COBIT (Control Objectives for Information and Related Technology) are examples of such regulations. Thus, the financial aspect and the strategic success of the data center organization in the long term depend heavily on compliance with such regulations. MORE introduces Business Performance Indicators (BPIs) as a subset of MORE KPIs measuring business oriented performance indicators regarding financial issues such as cost, utilization, and rejection rate. On other side, there are internal and external regulations that should be maintained, such as mean time between failures, mean time to repair, etc. SLA values such as response time, availability, and network throughput, etc. must be considered. Many factors influence each other in services described by SLAs, so the relationships need to be modeled and standardized. If these relationships are known, the interpretation of data obtained from a simulation model helps to get coherent approximations. The use of these relationships must be viewed critically. The simulation approach presented in MORE is based on techniques of discrete simulations, because individual interaction events between services and hardware can be defined as countable, finite sets (Banks 1998).

2 Concept: MORE in a Nutshell

The business objectives of deployment of BPIs as an important part of data center related KPIs are:

- To comply with national legislation and inter-governmental conventions
- To better internalize BPIs in local, intra-organizational activities, and decision-making processes
- To technically integrate BPIs in local IT environments and beyond
- To better analyze employability and impact of BPIs for selected case-specific business environments and business networks
- To technically access BPIs for a range of users, regardless the level of pre-existing knowledge in ICT skills
- To foster usage of BPIs based on the impact they generate in business, society, and environmental protection

MORE considered various perspectives in data center sector. These perspectives focus on three interrelated areas (Fig. 1): Environment, Usage Effectiveness, and

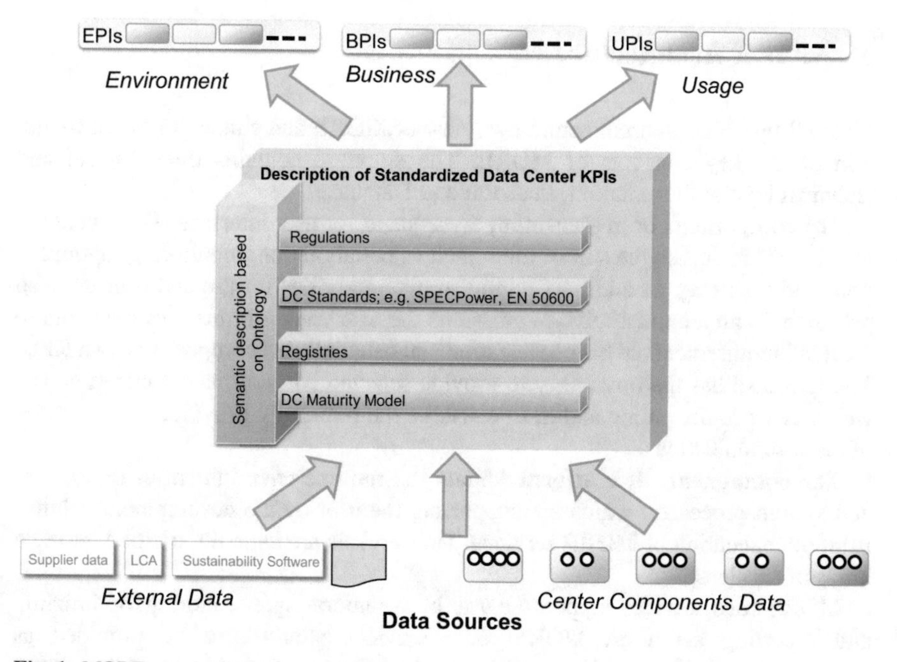

Fig. 1 MORE standardization

Business. They are represented and aggregated as data center Key Performance Indicators (KPIs) in the MORE concept and are divided into the following: Environmental Performance Indicators (EPIs), such as energy, water, CO_2 consumption, and other polluting emissions; Usage Performance Indicators (UPIs), such as Power Usage Effectiveness (PUE), Carbon Usage Effectiveness (CUE), and Water Usage Effectiveness (WUE), and Business Performance Indicators (BPIs), such as Cost and SLA Values including response time, availability, and network throughput, etc.

In MORE we define different components of a data center and how any changes to their values influence the optimization of energy consumption. The desired optimum should be set so that lower power consumption represents the trade-off between execution time (service level) and energy consumption of data center components. To obtain optimum power consumption, a trade-off must be made between execution time on a service level and the way components are utilized, such as turning off of some servers.

Optimization approach based on Non Deterministic Resource Framework (NDRF)—The NDRF evaluates approaches to saving energy in data centers. Furthermore, it includes a data center simulation that uses a defined input (service descriptions) and a characterization of hardware to provide a quantitative result, which includes energy saving potential and expected service levels. Thus NDRF establishes the correlation between software services and its corresponding consumption behavior on data center components like energy consumption on the server, consumption of the cooling, and storage (Zenker 2012).

3 MORE Architecture

Figure 2 provides an architecture overview of MORE and concludes the introduction of the key concepts of MORE. The structure contains three logical and technical layers: Presentation, Platform and Database.

The **components of presentation layer** stand for user interface (UI) where the user is able to access the suit of tools such as collaboration, monitoring, optimization, and reporting. In addition, administrative activities for the end-user are also provided. With administrative permissions the user can customize his own data to meet the requirements of his organization and calculate and compose his own KPIs. The user also has the possibility to provide selected KPIs of his organization as a web-service to the public and/or to consume the publically provided web-services of other organizations.

The components of Platform contain the runtime environment of the system and system processes, which are supporting the user by the development, administration, execution of MORE services. The services are logically divided into two kinds of services;

MORE suite services support the user by collaborating, monitoring, optimizing, and reporting activities. MORE suite services would also be provided as web-services which would be used by third party system (external systems, smart

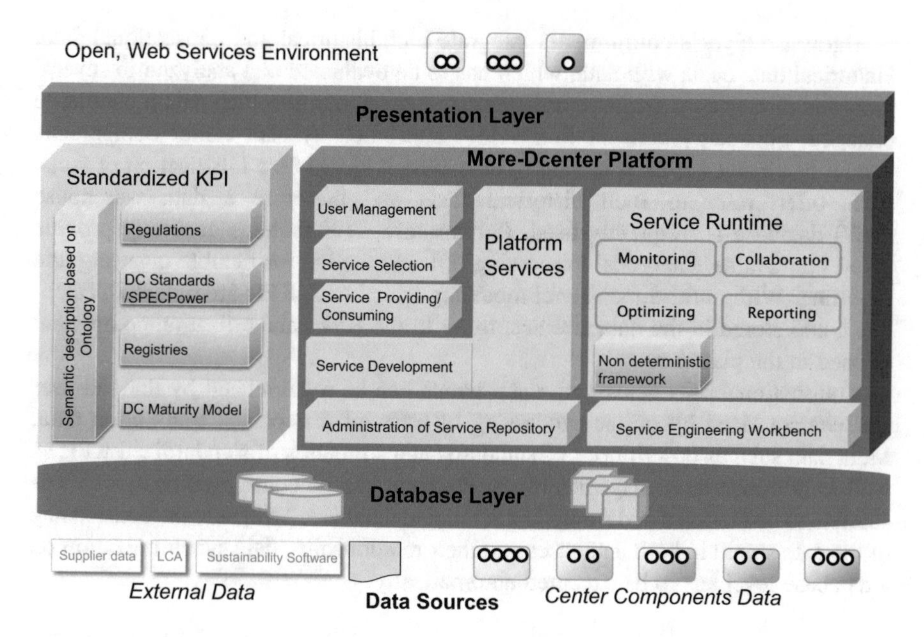

Fig. 2 MORE architecture

phones, PDAs, etc.). MORE should support the user by developing, calculating, composing, searching, providing, and consuming KPIs. Therefore, the platform supports the processing of the services from tow perspectives:

Service runtime is responsible for processing MORE suit services. The "Service Engineering Workbench" of MORE suites enables the developer to customize and manage MORE tools services and, Platform services are responsible for the management and processing of KPIs, and they support the administration of the KPI Service Repository.

All services processed by the MORE suite platform deal with KPIs, even optimizing activities based on monitoring of KPI values. The needed standardization of KPIs in the data center sector is defined by the KPI Standardization component which the semantic description approach (language) of KPI defines using ontology. The main contribution of the standardization component is based on an approach which guides one on how to drive data center KPIs with a unified format and characterization to ensure KPIs consistency. Data center KPI derivation is not only driven by national and international environmental standards, compliance frameworks, or data center related standards,[1] Data center Usage Effectiveness (power, water, and carbon), ITIL or COBIT, but also specific enterprise regulations in order to enable data center organization driven by its own specific KPIs. In this way, the system reaches high performance related flexibility.

[1] As DC Maturity Model, ISO 14040 Series Standard, Life Cycle Assessment (LCA), EN 50600, SPECPower, and SLAs.

Database layer's components integrate both historical and transactional data. Historical data deals with data, which has to be evaluated and analyzed in several facts and dimensions. Because of historical characterization such data it cannot be changed. So energy consumption values consumed by data center components cannot be changed from past values, otherwise it would be a falsification of facts. KPIs often refer to such historical facts and therefore a data warehouse (DW) database is frequently used. Furthermore, DW systems generally provide an ad hoc Online Analytical Processing (OLAP) engine for flexible querying and reporting. With multi-dimensional modeling techniques, KPIs are modeled, calculated, and stored in the database according to the KPI standardization component defined in the platform.

Transactional data deals with data, which can be manipulated by the end-user and characterized by online processing of relevant master and transaction data. Meta data such as descriptions of standards and structures of standardized KPIs as well as administrative activities of the Platform can be considered data, which is suitable to be modeled and stored in a relational database. To ensure the consistency of the data stored in both databases and their relationships, data synchronization on a database level has to be executed automatically.

3.1 Reporting

A major objective of the MORE is the identification of potential information of energy and other environmental data in data centers. This entails the collection, aggregation and providing of data in a standardized form. On one side, it serves various reporting needs in the company such as administrative reporting, mandatory corporate reporting, SLA reporting, corporate communication, and management reporting, and on the other hand, it facilitates and controls operational and strategic business decisions such as supplier selection of spare parts and energy supplier, as well as optimizing data center infrastructure, e.g. supplying power to different hardware components according to load balancing techniques, optimizing cooling and distribution of data storage, and controlling IT software service output from the data centers. The MORE Services help to access relevant internal/external energy and environmental information sources. The users then choose among consumption channels to access internal and external data.

Supply and distribution chains can be better supported by more reliable and faster ways to collect and process data for product specific EPDs "Environmental Product Declarations". EPDs are international tools used for transparent and reliable reporting of environmental aspects. They also report impacts of products and services in business to business transactions and business to customer transactions.

3.2 *Monitoring and Optimization*

As a basis for a reliable and conclusive interpretation of events and states of a data center, the monitoring component is important. Monitoring itself is important to identify problems like outages of server or service components, though it is nowadays widely used to support ITIL based processes like incident or problem management. As intended by MORE, the usage of monitoring data enables an online view of current system states with the goal of optimizing the current usage effectiveness of the whole data center.

This can only be enabled if both the structure of the data center's components (hardware and software) and the monitoring levels are corresponded and described semantically by descriptive standardized language (ontology). This ontology will be useful for different monitoring systems in order to enable the usage of a large variety of client systems. Especially the usage of open source components, like Nagios[2] (2012), is a major enhancement for MORE and the optimization of data centers.

MORE provides automatic usage adaptation to existing resources used by requested software services so that the energy consumption can be optimized in compliance with SLAs. The relationship between the energy consumption of resources and the loads of the requested services is usually a non-linear relationship. Non-deterministic methods are used to ensure appropriate automatic optimization of resources. MORE does the online forecast of future operations and their impact on the energy consumption of the data centers. This forecast is used in order to provide information about optimization means that include more than just the migration of services to other servers. In each case, co-hosting on a single node is considered in order to have effective resource utilization. In the same way, distributing on different node is considered to ensure compliance with existing SLAs.

The monitoring system provides all necessary data for the NDRF component in the architecture in order to predict future system states. This prediction is used by an IT-Operator (can by automatic) to optimize the current utilization of the complete data center or provide recommendations for manual infrastructure adapting activities (Fig. 3).

The basis of each prediction is an assumption that is processed by the NDRF, and after a positive evaluation of this assumption it becomes a likely candidate which will be used for further processing. After the confirmation phase of that candidate and subsequent validation of its possible outcomes is presented to the MORE framework presentation layer. This set of assumptions is not limited to new entries. The set will be filled with standard-procedures in IT Operations, like migration of virtual machines, power-down of hardware components, etc. Each new optimization approach may create promising assumptions that are created to the set of possible assumptions. As review cycles and information validation is important

[2] Nagios is an open source computer system monitor, network monitoring and infrastructure monitoring software application.

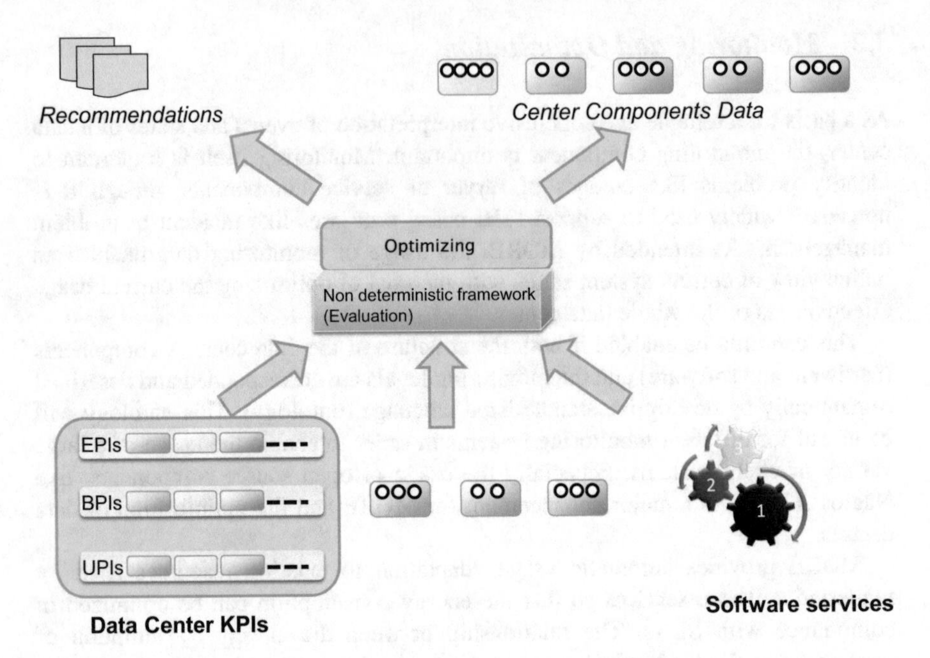

Fig. 3 MORE optimization

for the outcome of the optimization this process is also standardized. It will follow action research principle. Figure 4 represents the basic set up of the upcoming optimization challenges and the stabilization of each result.

3.3 *Collaboration*

The MORE collaboration platform promotes the integration of data center organization in smart grid infrastructures of urban regions by sharing selected EPIs and KPIs as services and integrating task holders into the platform (data center partner, equipment supplier, power supplier, IT service provider, etc.) to promote collaboration between them for smart grids in the form of broadband infrastructure including sharing engineering works and reusing infrastructures and services.

Data centers can also benchmark themselves against others. In this way they can ensure their efficiency (policy) meets the markets requirements. Data center KPIs as services are provided by MORE for benchmarking many organizations especially SMEs that don't have the ability to asses all data center components or the knowledge. Approximations and specifications of data shared by other members through the collaborative network platform can be used when they show the same components with similar structure. Average values can also be selected from supplier database.

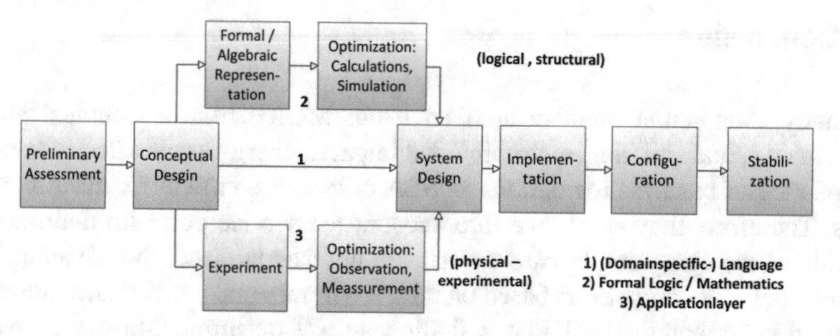

Fig. 4 Basic set up of stabilized and validated data according to Ortner (1999)

MORE also establishes a collaborative network platform to promote the collaboration between involved organizations through exchanging of expertise and information. For non-structural data, MORE provides collaboration tools as Document Management Systems (DMSs), internet forums, and links to social networks. Depending on the KPIs provided by MORE, many strategic and operational activities can be supported, such as optimizing cloud computing processes by using external data center resources of partner organizations, supporting decisions related to purchasing new equipment, or establishment new eco friendly technologies. These are all examples of results of such collaborations.

3.3.1 Collaboration Example

Based on the analysis of KPIs provided by MORE monitoring and reporting tools, the management of an organization instructs the data center administrators to analyze the energy market in order to find possible power suppliers that are the most eco friendly, inexpensive, reliable, and integrated with smart grid city infrastructure. The data center staff analyzes the KPI services provided in the MORE collaboration platform by different power supplier networks. According to EPIs of the supplier and the requirements of the management, the staff can select power suppliers quickly and efficiently. They can also buy power from different suppliers for different periods of time to ensure the best possible performance in the data center. Thus, power used at night can be delivered by a specific supplier with more favorable terms, such as price and better EPI values. KPIs are not the only deciding factors supporting such decisions, experience of other organizations and their reviews of the power supplier are as well, e.g. reliability or supporting of services. MORE considers evaluations of other organizations and provides unstructured information, such as documents about pricing models and contract terms, creating optimal conditions to make a decision.

4 Conclusion

Normally, data center member staff envisions MORE from a technical point of view. In an actual situation, environmental impacts, energy usage effectiveness, and compliance of business regulations as SLA criteria are critical for customer decisions. Therefore, they must take into account the service portfolio definition and offering of the company. In case criteria are not standardized, the advantages and disadvantages of competitors based on their environmental performance are neither measurable nor applicable. KPIs[3] will allow an ICT department to simply evaluate the advantages of a competitor and to incorporate them into the local IT environment through MORE services. Resulting in an assessment of competitive advantages, KPIs can then be offered uniformly by all providers. This case assumes that the company wants to add a KPI to its service portfolio that is not included in their services. The company assesses theoretical information about this KPI, obtains the internal information to deploy it, and the ICT department incorporates the KPI into the local IT environment. As outlined in the architectural model, data center member staff use their technical knowledge to retrieve KPI databases and third party services. Based on, MORE-platform provides streamlined access to the KPI. The next phase of MORE is to deploy it in the real market to evaluate its capability.

References

Austrian Standards Institute (2012) Information technology—data centre facilities and infrastructures. https://www.astandis.at/shopV5/Preview.action;jsessionid=F6355B3CE70A679DC5ED41B6F6BD3D99?preview=&dokkey=410276&selectedLocale=en. Accessed 1 Feb 2015

Banks J (1998) Handbook of simulation: principles, methodology, advances, applications, and practice. Wiley, New York

Beims M (2009) IT Service Management in der Praxis mit ITIL 3: Zielfindung, Methoden, Realisierung. Carl Hanser Verlag, Munich

Belady C, Azevedo D, Patterson M, Pouchet J, Tipley R (2010) Carbon usage effectiveness (CUE): a green grid data center sustainability metric

Federal Environment Agency (2012) Timely replacement of a notebook under consideration of environmental aspects. https://www.umweltbundesamt.de/sites/default/files/medien/461/publikationen/4317.pdf. Accessed 1 Feb 2015

Naigos (2012) Naigos core overview. http://nagios.sourceforge.net/docs/3_0/about.html. Accessed 1 Feb 2015

Ortner E (1999) Konsequenzen einer konstruktivistischen Grundsatzposition für die Forschung in der Wirtschaftsinformatik. In: Schütte R, Siedentopf J, Zelewski S (eds) Wirtschaftsinformatik und Wissenschaftstheorie Grundposition und Theoriekerne. Institut für Produktion und Industrielles Informationsmanagement, Essen

SPEC (2012) SPEC—power and performance. http://www.spec.org/power/docs/SPECpower_ssj2008-Design_ccs.pdf. Accessed 1 Feb 2015

[3] Only data center KPIs that contains information for public.

The Green Grid (2008) Green grid data centre power efficiency metrics: PUE & DCIE. http://www.eni.com/green-data-center/it_IT/static/pdf/Green_Grid_DC.pdf. Accessed 1 Feb 2015

The Green Grid (2011) Data center maturity model. http://www.thegreengrid.org/~/media/WhitePapers/Data%20Center%20Maturity%20Model%20White%20Paper_final.ashx?lang=en. Accessed 1 Feb 2015

Zenker N (2012) Bestimmung des Energiebedarfs von Rechenzentren—non-deterministic resource framework. Verlag Dr. Kovac, Hamburg

Using Social Media to Improve Environmental Awareness in Higher Education Institutions

Thabo Tlebere, Brenda Scholtz, and André P. Calitz

1 Introduction

Environmental awareness campaigns provide individuals with the necessary knowledge, skills and attitude to prevent issues facing the environment (Apil and Okaka 2013; Talero 2004). Awareness of environmental issues is important since it can assist individuals and social groups to acquire a sense of sympathy towards the environment (Zsóka et al. 2013). Social media are focused on social aspects such as communities, participation, openness, conversations and connectedness (Katajisto 2010). Several environmental awareness campaigns have been successfully conducted using social media (Idumange 2012; Kaur 2015). According to Kriek (2011) social media have become very popular over the past years. The number of adults who engage on them has increased by 57 % from 2007 to 2011. Research related to the design of social media campaigns is limited especially in the context of environmental issues in higher education. In addition there is a need for evaluating the social media used in these campaigns.

The primary purpose of this paper is to investigate whether environmental awareness campaigns have the ability to improve peoples' environmental knowledge and ultimately their awareness. In order to achieve this, a conceptual Social Media for EnviroNmental Awareness (SMENA) model was designed. The model was implemented in a South African university to facilitate an environmental awareness campaign that seeks to increase the environmental knowledge and awareness of students. The SMENA incorporates and uses various popular social media namely Twitter, Facebook and an in-house social media website (SMENA website) to create awareness of environmental issues. The participants were

T. Tlebere • B. Scholtz (✉) • A.P. Calitz
Department of Computing Science, Nelson Mandela Metropolitan University (NMMU), Port Elizabeth, South Africa
e-mail: brenda.scholtz@nmmu.ac.za; andre.calitz@nmmu.ac.za

J. Marx Gómez, B. Scholtz (eds.), *Information Technology in Environmental Engineering*, Springer Proceedings in Business and Economics,
DOI 10.1007/978-3-319-25153-0_9

required to use the SMENA social media to perform activities as requested by the researcher. The environmental knowledge of the participants was tested before and after the participants performed tasks on the SMENA social media as part of the environmental awareness campaign.

A literature review regarding social media used for environmental awareness was conducted (Sect. 2). There are several evaluation factors and guidelines for improving the success of social media campaigns (Sect. 3). The research methodology used was selected in order to meet the research objectives of this study (Sect. 4). The participants in the study were undergraduate students and whilst the results were generally positive, some challenges were encountered (Sect. 5). An analysis of the results revealed several conclusions and recommendations (Sect. 6).

2 Social Media for Environmental Awareness

Environmental education is associated with providing individuals with an in-depth understanding of environmental issues and also with the provision of skills which can assist individuals to make accurate decisions with regard to the environment (EPA 2013). Environmental awareness is a component of environmental education. According to Hungerford and Volk (1990) environmental awareness provides social groups and individuals with awareness and sensitivity towards the environment and the issues affecting it. Altaher (2013) confirms that environmental awareness is the perception of human activities affecting the environment and a behavioural inclination to protect the environment. Furthermore, there is a positive correlation between environmental awareness and environmental knowledge which implies that an increase in environmental knowledge leads to an increase in environmental awareness (Hungerford and Volk 1990; Ramsey and Rickson 1976; Zsóka et al. 2013). One objective of higher education institutions is to adopt environmentally friendly practices and to improve environmental awareness of individuals within the environment of the institution (Disterheft et al. 2012).

Chaineux and Charlier (1999) argue that people's awareness of the environment can be improved if seamless communication is maintained between environmental information providers and consumers. The interaction between the provider and the consumer should incorporate information that can enhance the consumer's environmental knowledge, skills and knowledge about technologies that can support the environment. The information should also include ways to identify and access environmental information and information sources. Lively (2011) argues that social media have the ability to be the driver of sustainable development and can assist in creating environmental awareness. Social media applications enable users to interact with each other from online communities (Shinton 2012). Social media users interact by generating and distributing content (media and text), which enables them to communicate and collaborate with each other (Jussila et al. 2011; Lee and Ma 2012).

Social media can be divided into different categories (Katajisto 2010). These categories are: content creating and publishing in social media; content sharing in social media; collaborative producing in social media; virtual worlds and social networks. Social networks are the most popular types of web-based social media (Cheng et al. 2010). Facebook is the most popular social networking site as it has more than one billion users who are active on it on a monthly basis (Khang et al. 2012; Lee 2012). Other popular social media include the video sharing channel YouTube with 35 h of video content posted every minute, Twitter, a micro-blogging site with 200 million users, and Flickr which made more than six billion photographs available to users (Khang et al. 2012; Madrigal 2013).

Environmental campaigns rely heavily on information and seek to provide individuals with the proper knowledge, skills and attitude to address environmental problems (Apil and Okaka 2013; Mooney et al. 2009). Mooney et al. (2009) confirm that environmental awareness campaigns play a big role in improving an individual's environmental awareness and changing their attitudes towards environmental issues. Numerous environmental campaigns have been conducted using social media, for example social media have been used to foster environmental behaviour, to get petitions signed, to provide news, to provide motivation and to improve awareness (Kaur 2015). Social media are effective in carrying out environmental campaigns because they have the ability to distribute information quickly and are cost effective. Idumange (2012) confirms that social media are able to support environmental initiatives because they are able to reach a wide spectrum of audience. They can be easily accessed; they are easy to use; they enable users to get instant responses, and they allow instant modification of responses by means of comments. Idumange (2012) also adds that the various social media sites such as Blogs, Facebook, Twitter and YouTube can be used in multiple ways to raise environmental awareness. However, there is a cost associated with using these social media for environmental campaigns, which is the cost of the time and effort required for participation in content creation and sharing.

3 Social Media Usage

The use of social media sites increases daily and their value is mainly dependent on user interactions (Xu et al. 2012). Social media can be used for both leisure and professional purposes (Monnonen and Runonen 2008). Within organisations marketing professionals utilise social media to conduct marketing campaigns in order to support organisational brands and to improve relations with stakeholders (Khang et al. 2012; Michaelidou et al. 2011). In higher education the use of social media is common amongst university students who use them during their leisure time (Dhume et al. 2012) and are also accessed at university where they are used by students for communication, collaboration and learning (Tess 2013). The use of social media for learning in higher education institutions makes learning more student-centred (Gikas and Grant 2013). Tess (2013) adds that including social media in learning in higher education can improve the effectiveness of communication. However, Tess (2013) agrees that most students that have social media accounts rarely use social media for educational purposes; instead they mostly use it for communicating with friends and career networking. Furthermore, Hussain (2012) confirms that students mainly use social media for enjoyment, accessing their academic information and current affairs.

Critical Success Factors (CSFs) and guidelines were identified for using social media within organisations and higher education institutions for various purposes such as communication and marketing (Table 1). One of these factors is internet usage, since policies relating to internet usage can affect students' use of social media (Falahah and Rosmala 2012). The background of the user and their social media usage behaviour can also affect their usage. The goals and objectives of the campaign and the social media should be identified in the beginning phase of the campaign (Roberts 2010; Thackeray et al. 2008; Walther 2010). It is also important to select social media applications that best match the campaign's objectives and that allow students to carry out work in ways that are common to them (Roberts 2010; Thackeray et al. 2008). Roberts (2010) suggests that students' readiness and a change management strategy should be considered prior to starting the campaign. It is important to identify metrics that will determine the value of using social media for the campaign (Roberts 2010) and to decide on the campaign duration (Thackeray et al. 2008). During campaign implementation frequent updates on the social

Table 1 Critical success factors and guidelines for social media awareness campaigns

Factors	Guidelines	Purpose
Internet usage	Identify the university policies for Internet usage and access (Falahah and Rosmala 2012)	Higher education institutions
User background	Determine the background and social media usage behaviour of users (Falahah and Rosmala 2012)	
Goals and objectives	Identify the goals and objectives for using social media and closely match them to the goals and objectives of the campaign (Roberts 2010; Thackeray et al. 2008; Walther 2010)	Communication and marketing
Social media selection	Select social media applications that best match the campaign's objectives and that allow students to carry out work in ways that are common to them (Roberts 2010; Thackeray et al. 2008)	
Readiness and change management	Evaluate students' readiness and construct a change management strategy (Roberts 2010)	Communication
Value metrics	Identify metrics that will determine the value of using social media for the campaign (Roberts 2010)	
Duration of programme	Identify the duration of the campaign (Thackeray et al. 2008)	Marketing
Frequent information updates	Provide frequent updates on the social media site (Walther 2010)	
Involvement of users	Allow users to get involved and give a reason why they should participate (Walther 2010)	
Authenticity	Conduct the campaign with honesty and authenticity (Walther 2010)	

media site should be provided (Walther 2010). Allow users to get involved and give a reason why they should participate. Conduct the campaign with honesty and authenticity.

4 Research Objectives and Methodology

The Social Media for EnviroNmental Awareness (SMENA) conceptual model (Fig. 1) was designed to facilitate the design and implementation of an environmental awareness social media campaign. The main aim of the SMENA model is to improve peoples' awareness of environmental issues. The SMENA model lifecycle is divided into three different phases, which are: the strategy development phase, planning phase and the implementation phase. The model takes into consideration the factors and guidelines for using social media campaigns (Table 1). A field experiment approach was adopted for this study over a 4-week period (Saunders et al. 2009). The participants of the study were undergraduate students studying Information Systems or Computer Science degrees at a South African university.

In the strategy development phase the environmental goals and objectives were identified and prioritised to two environmental categories per week (so eight in total). A survey of participants was undertaken to determine the user background. In the planning phase the preferences of students with regards to social media was identified and Twitter and Facebook were identified as the most popular. In addition, an environmental knowledge pre-test evaluation was also conducted with the participants to determine their current level of environmental knowledge before being exposed to environmental information. In the implementation phase frequent information updates were provided as recommended by Walther (2010). The administrator focused on two environmental categories per week and posted information (Blogs and videos) related to these categories on the SMENA social media throughout the week (Fig. 2). This was repeated each week and information

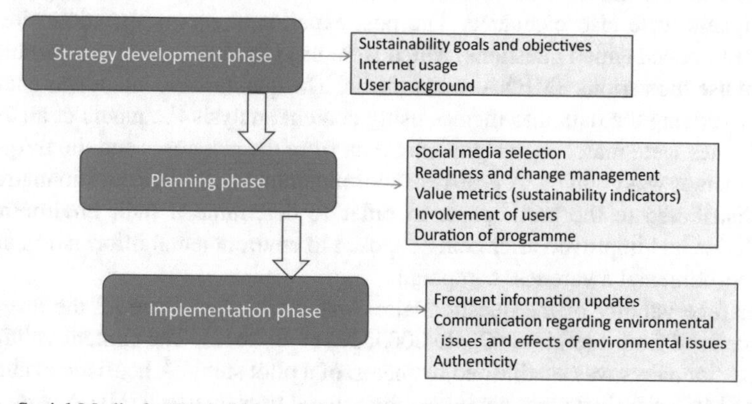

Fig. 1 Social Media for EnviroNmental Awareness (SMENA) conceptual model

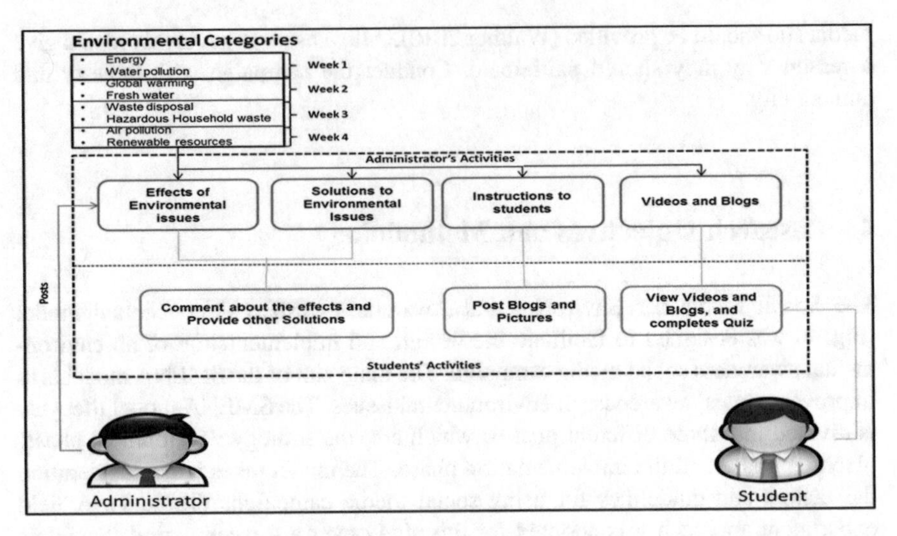

Fig. 2 SMENA social media activities

included effects of environmental issues related to each category and solutions that can be implemented to prevent these environmental issues. The participants were requested to provide their opinions by posting comments regarding the effects of these environmental issues and to provide other solutions that can be implemented to eradicate these environmental issues. The administrator also posted instructions requesting the participants to post blogs and pictures about environmental concerns of their surroundings. The participants were also requested to watch videos and read blogs based on the environmental categories and to complete a quiz that was directly based on the videos and blogs. These activities were conducted on all the SMENA social media used for the experiment.

A post-experiment evaluation was conducted at the end of the 4-week period of the experiment. The usability of the SMENA website and the extent to which the SMENA social media were able to improve environmental knowledge of the participants were also evaluated. The post-experiment survey questionnaire also included opened ended questions, which were used to identify reasons participants did not use the various SMENA social media. The qualitative content was analysed by categorising the data into themes using content analysis (Saunders et al. 2009). The themes were matched to significant data from the responses and the frequency of the themes was counted. A post-test environmental knowledge questionnaire was also distributed to the participants in order to determine if their environmental knowledge had improved after being exposed to environmental information during the environmental awareness campaign.

The face validity of the questionnaires was established since all the questions were derived from literature (Coyle 2005; Xu et al. 2012). The content validity of the questionnaires was established by means of a pilot study. A heuristic evaluation of the SMENA website was conducted by several user interface (UI) experts using

Nielsen's heuristics (Nielson 1995). The recommendations provided by the experts were used to improve the usability of the website.

5 Analysis of Results

5.1 Participant Profile

The participants were 72 undergraduate students studying towards a degree in Computer Science and/or Information Systems, with the majority (67 %) registered for a science degree (Table 2). The majority of participants were male (81 %) and most them were between the age of 18 and 25 (97 %). Over half (53 %) of them were English speaking.

5.2 Research Instruments

The research instruments used in the study were a pre-test questionnaire and a post-test questionnaire which were used to evaluate (1) the reasons for not using the SMENA social media (post-test); and (2) the environmental knowledge of the participants (pre- and post-test). The SMENA social media consisted of the SMENA website, the SMENA Facebook page and the SMENA Twitter page. Several themes were extracted from the qualitative data collected from the questionnaires. The most frequent theme which was identified regarding the reasons for not using the SMENA website is "*It is time consuming*" (f = 7) (Table 3). Other frequent themes which were identified include "*Not interested*" (f = 2) and "*Videos took long to stream and download*" (f = 2).

The most frequently identified theme related to reasons for not using the SMENA Facebook page was "*I do not use Facebook frequently*" (f = 5) (Table 4). The second most frequently identified themes was "*Information not interesting*"

Table 2 Participant profile (n = 72)

	n	%
Gender		
Male	58	81
Female	14	19
Total	**72**	**100**
Home language		
English	38	53
Afrikaans	13	18
Xhosa	12	17
Other	9	12
Total	**72**	**100**

Table 3 Reasons for not using the SMENA website (n = 15)

Theme (SMENA website)	f	Examples of actual response
It is time consuming	7	I got caught up in other subjects and assignments
Not interested	2	I'm not interested to waste my little time bit of free time on the website
Videos took long to stream and download	2	One of the times I tried to view a video it did not work so I left it

Table 4 Reasons for not using SMENA Facebook (n = 19)

Theme (SMENA Facebook) page	f	Sample response
Do not use Facebook frequently	5	I hardly use Facebook
Information not interesting	2	The articles posted could be more interesting
It is time consuming	2	Time constraints
Theme (SMENA Twitter)	f	Examples of actual response
I do not use Twitter	8	No Twitter account, not interested in twitter
I do not like Twitter	2	Don't like or use twitter

(f = 2) and "*It is time consuming*" (f = 2). The most frequent theme that was identified regarding the SMENA Twitter page is "*I do not use Twitter*" (f = 8). The second most frequent theme is "*I do not like Twitter*" (f = 2).

Furthermore, some other negative comments made by participants were recorded during the campaign. Some participants were negative towards the campaign since it increased their workload but was not a direct part of their course curriculum and did not contribute significantly to a mark for their course. Some participants reported that the campaign "*is a waste of time since participation will not contribute towards our final class mark*". Another challenge identified was the fact that social media is generally prohibited in the computer laboratories and this also had a negative effect on the attitude of students. The pre- and post-environmental knowledge questionnaire consisted of 16 Environmental Knowledge Questions (EKQs) with 2 questions for each environmental category. For example the questionnaire includes two questions which are focused on energy. The question that scored the highest score (87 %) in the pre-test environmental knowledge evaluation is EKQ14 which is in the renewable energy category (Fig. 3). Furthermore, EKQ3 which is in the water pollution category scored the lowest score (0 %). The average percentage scored by the participants was 53.3 %.

In the post-test environmental knowledge evaluation EKQ14 had the highest score (94.2 %) (Fig. 4). The score for EKQ3 improved from 0 to 29 %. The question that scored the lowest score (8.7 %) in this evaluation is EKQ8 which is within the fresh water category. Furthermore, the scores achieved by EKQ8 (8.7 %) and EKQ1 (78.3 %) remained the same in both evaluations. The scores of all other questions increased from pre-to post evaluation therefore there was an improvement in the overall average percentage score to 63.9 %. This implies that the participants environmental knowledge improved by 10.6 %.

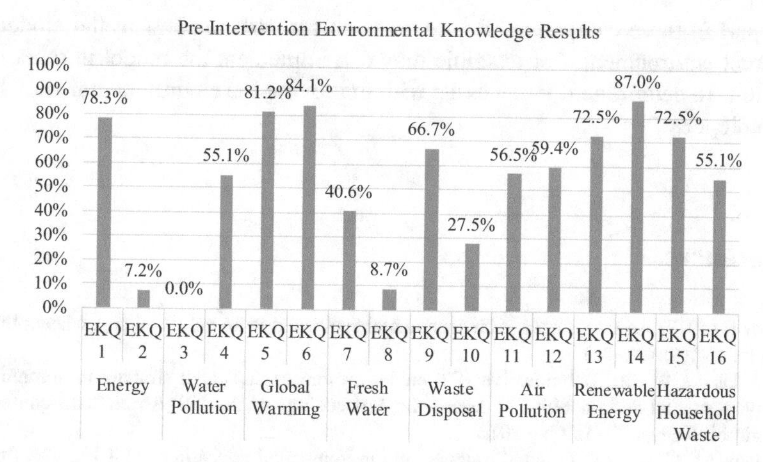

Fig. 3 Pre-intervention environmental knowledge results (n = 72)

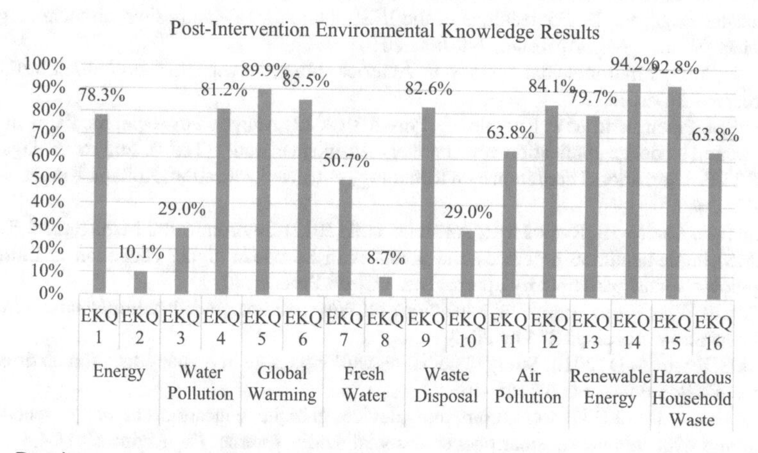

Fig. 4 Post-intervention environmental knowledge results (n = 72)

6 Conclusions and Future Work

Several key contributions are made by this paper for the social media research community as well as for environmental education researchers. The SMENA was successful in improving the environmental knowledge of students in higher education institutions. Challenges were encountered relating to the slow Internet speed, negative attitudes towards the study and time constraints of course demands. Future researchers could implement and evaluate the SMENA at a university with more senior students (postgraduate students). The fact that students mostly use social media to acquire academic oriented information can be used as the basis to argue that the application can be successful in achieving environmental goals if it is included in the university curriculum and students are incentivised in some way to take part in these campaigns. Future research can also implement the model in a different environment. For example they can implement the model in other organisations to determine if it can assist with improving the environmental knowledge of employees.

References

Altaher H (2013) An assessment of environmental awareness in an industrial city. Manag Environ Q Int J 24(4):422–451

Apil J, Okaka W (2013) Innovative ICT public awareness campaign strategy to communicate environmental sustainability in Africa. In: Proceedings of the IST-Africa 2013 conference, Nairobi, Kenya, 29–31 May 2013

Chaineux M, Charlier R (1999) Strategies in environmental education. Int J Environ Stud 56 (6):889–905

Cheng E, Davis S, Burnett I, Ritz C (2010) The role of experts in social media—are the tertiary educated engaged? In: Proceedings of the IEEE international symposium on technology and society, Wollongong, Australia, 7–9 June 2010

Coyle K (2005) Environmental literacy in America. http://www.neefusa.org/pdf/ELR2005.pdf. Accessed 2 Dec 2012

Dhume SM, Pattanshetti MY, Kamble SS, Prasad T (2012) Adoption of social media by business education students: application of technology acceptance model (TAM). In: Proceedings of the 2012 I.E. international conference on technology enhanced education, Kollam, Kerala, 3–5 Jan 2012

Disterheft A, Caeiro SS, Ramos MR, Azeiteiro UM (2012) Environmental Management Systems (EMS) implementation processes and practices in European higher education institutions—top-down versus participatory approaches. J Clean Prod 31:80–90

EPA (2013) What is environmental education? http://www2.epa.gov/education/what-environmental-education. Accessed 23 Oct 2013

Falahah S, Rosmala D (2012) Study of social networking usage in higher education environment. Procedia Soc Behav Sci 67:156–166

Gikas J, Grant MM (2013) Mobile computing devices in higher education: student perspectives on learning with cellphones, smartphones & social media. Internet High Educ 19:18–26

Hungerford H, Volk T (1990) Changing learner behavior through environmental education. J Environ Educ 21(3):8–22

Hussain I (2012) A study to evaluate the social media trends among university students. Procedia Soc Behav Sci 64:639–645

Idumange J (2012) The social media as a platform for creating environmental awareness in the Niger Delta region. http://www.thenigerianvoice.com/news/99332/50/the-social-media-as-a-platform-for-creating-enviro.html. Accessed 10 June 2013

Jussila J, Kärkkäinen H, Leino M (2011) Benefits of social media in business-to-business customer interface in innovation. In: Proceedings of the 15th international academic MindTrek conference: envisioning future media environments, Tampere, Finland, 28–30 Sept 2011

Katajisto L (2010) Implementing social media in technical communication. In: Proceedings of the 2010 I.E. international professional communication conference, Enschede, Netherlands, 7–9 July 2010

Kaur K (2015) Social media creating digital environmental publics: case of Lynas Malaysia. Public Relat Rev 41(2):311–314

Khang H, Ki E, Ye L (2012) Social media research in advertising, communication, marketing, and public relations, 1997–2010. J Mass Commun Q 89(2):279–298

Kriek L (2011) Mobile social media for a private higher education institution in South Africa. In: Proceedings of the 2011 3rd symposium on Web Society, Port Elizabeth, South Africa, 26–28 Oct 2011

Lee D (2012) Facebook surpasses one billion users as it tempts new markets. http://www.bbc.co.uk/news/technology-19816709. Accessed 20 May 2013

Lee CS, Ma L (2012) News sharing in social media: the effect of gratifications and prior experience. Comput Hum Behav 28(2):331–339

Lively A (2011) Is social media the missing key in urban sustainability? http://eucenterillinois.blogspot.com/2011/11/is-social-media-missing-key-in-urban.html. Accessed 15 Apr 2012

Madrigal A (2013) Chart: where Yahoo's Tumblr ranks next to Twitter, Instagram, and Pinterest. http://www.theatlantic.com/technology/archive/2013/05/where-yahoos-tumblr-ranks-next-to-twitter-instagram-and-pinterest/276017/. Accessed 20 May 2013

Michaelidou N, Siamagka NT, Christodoulides G (2011) Usage, barriers and measurement of social media marketing: an exploratory investigation of small and medium B2B brands. Ind Mark Manag 40(7):1153–1159

Monnonen P, Runonen M (2008) SMEs in social media. In: Karahasanovic A (ed) New approaches to requirements elicitation & how can HCI improve social media development? Proceedings of the NordiCHI 2008 workshop, Lund, Sweden, Oct 2008. Tapir Academic Press, Trondheim, pp 85–90

Mooney P, Winstanley A, Corcoran P (2009) Evaluating Twitter for use in environmental awareness campaigns. In: Proceedings of the China-Ireland information and communications technologies conference, University of Maynooth, Ireland, 19–21 Aug 2009

Nielson J (1995) 10 Usability heuristics for user interface design. http://www.nngroup.com/articles/113-design-guidelines-homepage-usability/. Accessed 19 June 2013

Ramsey CE, Rickson RE (1976) Environmental knowledge and attitudes. J Environ Educ 8 (1):10–18

Roberts S (2010) Critical success factors for enterprise social networking: four tips for ensuring the success of collaboration in your organization. http://resources.moxiesoft.com/success-factors-ty.html. Accessed 8 Oct 2012

Saunders M, Lewis P, Thornhill A (2009) Research methods for business studies, 5th edn. Pearson Education, Harlow

Shinton S (2012) #Betterconnected—a perspective on social media. Anal Bioanal Chem 402 (6):1987–1989

Talero G (2004) Environmental education and public awareness. http://worldfish.org/PPA/PDFs/Semi-Annual%20II%20English/2nd%20s.a.%20eng_F2.pdf. Accessed 25 July 2013

Tess PA (2013) The role of social media in higher education classes (real and virtual)—a literature review. Comput Hum Behav 29(5):A60–A68

Thackeray R, Neiger BL, Hanson CL, McKenzie JF (2008) Enhancing promotional strategies within social marketing programs: use of Web 2.0 social media. Health Promot Pract 9 (4):338–343

Walther D (2010) Critical success factors of social media marketing campaigns for consumer goods knowledge management. http://daenu.net/joomla/index.php?option=com_docman&task=doc_download&gid=40. Accessed 10 Aug 2012

Xu C, Ryan S, Prybutok V, Wen C (2012) It is not for fun: an examination of social network site usage. Inf Manag 49(5):210–217

Zsóka Á, Szerényi ZM, Széchy A, Kocsis T (2013) Greening due to environmental education? Environmental knowledge, attitudes, consumer behavior and everyday pro-environmental activities of Hungarian high school and university students. J Clean Prod 48:126–138

Sustainability Reporting by South African Higher Education Institutions

André P. Calitz, Margaret D.M. Cullen, and Samuel Bosire

1 Introduction

Sustainability has gained importance internationally, as is indicated in United Nations publications, such as the Global Compact and Principles of Responsible Investment (IoD 2009). According to the Brundtland Report, published by the United Nations World Commission on Environmental Development (WCED), sustainable development is defined as "development that meets the needs of the present without compromising the abilities of future generations to meet their own needs" (WCED 1987). Organisations are increasingly being challenged by sustainable development issues and are required to account for the consequences of their activities on the environment to society (Dimitrov and Davey 2011). Reporting is of help to communicate the activities of organisations.

Organisational activities and achievements should be reported from a holistic perspective and the Global Reporting Initiative (GRI)'s guidelines have been cited as a best practice reporting framework (Fonseca et al. 2011; Smith and Scharicz 2011). The GRI presents performance indicators from the economic, environmental, financial and social responsibility perspective (Microsoft Dynamics 2010).

A Public Higher Education Institution is established to serve the common good of society and therefore Higher Education Institutions (HEIs) should not only educate society about the importance of sustainability but also demonstrate sustainability practices in their operations. To this end, universities should lead by example by putting into practice the sound principles of Sustainability Reporting. The stakeholders in Higher Education have a legitimate claim on information about the operations of universities—especially those funded from public coffers.

A.P. Calitz (✉) • M.D.M. Cullen • S. Bosire
Nelson Mandela Metropolitan University, Port Elizabeth, South Africa
e-mail: andre.calitz@nmmu.ac.za; margaret.cullen@nmmu.ac.za; samuel.bosire@nmmu.ac.za

J. Marx Gómez, B. Scholtz (eds.), *Information Technology in Environmental Engineering*, Springer Proceedings in Business and Economics,
DOI 10.1007/978-3-319-25153-0_10

HEIs should assume leadership in sustainability and at the same time act as drivers of change towards a sustainable world as envisaged through declarations, charters and partnerships for sustainable development (Lozano et al. 2011). In light of the mandate of Higher Education, Stephens and Graham (2010) call on HEIs to take a lead in the transition to a more sustainable society by adopting sustainability practices—including reporting.

The process of developing sustainability reports is complicated by the multidimensional nature of these reports. Information required by these reports is often stored across multiple disparate systems and databases. A reporting tool that can integrate this data to produce one integrated sustainability report can greatly enhance the sustainability reporting process. This paper investigates the extent to which HEIs in South Africa implement sustainability reporting in their institutions.

The layout of the paper is as follows: Sect. 2 states the research objectives and methodology of this study. Section 3 highlights sustainability reporting and HEIs sustainability reporting. The results of an investigation into the sustainability reporting are presented in Sect. 4. Section 5 provides recommendations from the results of the literature and survey and discusses the conclusions.

2 Research Objectives and Methodology

This study consisted of a thorough literature review that provided a better understanding of sustainability reporting in HEIs. A combination of deductive and inductive approaches was followed because there is a growing body of literature on sustainability reporting in general however limited research has been conducted regarding the integrated sustainability reporting practices in Higher Education Institutions in South Africa.

In this study, a self-administered paper based questionnaire was sent to individuals responsible for the management of information at all South African universities. The *Sustainability Reporting Practices in Higher Education in South Africa* (*SRPHESA*) survey was completed by Chief Information Officers, Directors of Information Technology and managers responsible for management of information at all South African public universities. The questionnaire was developed from literature.

The questionnaire consisted of open-ended questions, as well as closed-ended questions in which a 5-point Likert scale was applied. The results of the questionnaire were statistically analysed to measure the intended outcomes. Questions in the survey were grouped into the following categories in concert with the themes of the research questions and objectives:

1. Higher Education stakeholders and their information needs
2. Monitoring of strategic plans in Higher Education
3. Institutional plans and Sustainability Reporting in Higher Education

4. Information culture in institutions and use of Business Intelligence (BI) tools and techniques
5. Elements of Sustainability Reporting

3 Sustainability Reporting in HEIs

Organisations produce different types of reports for various purposes. From an initial focus on environmental stewardship, steered by special interest groups, Sustainability Reporting is now prominent on the global agenda. A survey on sustainability indicates growth in Sustainability Reporting adoption (KPMG 2011). The survey's findings indicated the following:

- 95 % of the 250 largest global companies report on sustainability
- The highest reporting rates are associated with European organisations although North America and emerging markets continue to register phenomenal growth
- The rate of adoption of reporting varies across economic sectors
- Publicly traded companies outperform family-type organisations in embracing Sustainability Reporting

The large number of corporate failures in the last decade has prompted questions about the adequacy and relevance of traditional financial reports (Chen 2011; Hazelton and Haigh 2010; IoD 2009). Organisations have traditionally relied on financial reports to assess performance. However, today, stakeholders are increasingly demanding information regarding the performance of organisations—from various dimensions—in order to make informed assessments (Herzig and Godemann 2010; IoD 2009). Organisational reporting has matured in response to changing reporting requirements of stakeholders.

Factors such as the level of sophistication of an organisation's information systems, increasing demand for information by stakeholders and availability of reporting standards play a role in promoting organisational reporting. In addition, the role of professional bodies such as auditors, in attesting to the reliability of reported information as well as the growing recognition of the importance of holistic reporting by organisations, also contributes to emphasising Sustainability Reporting.

Internationally, the Global Reporting Initiative (GRI) has emerged as a generic global benchmark for reporting on sustainability (Dumay et al. 2010). In South Africa, the King III Report on corporate governance has given impetus to the adoption of Sustainability Reporting (IoD 2009).

According to Microsoft's Environmental Sustainability White Paper (Microsoft Dynamics 2010), pressure from regulatory bodies and the media, coupled with more rigorous investment criteria that include sustainability, have contributed to accelerating the need for Sustainability Reporting solutions. The Microsoft White Paper (Microsoft Dynamics 2010) also alludes to the urge by organisations to enhance their reputation and public standing by adopting Sustainability Reporting practices.

Petrini and Pozzebon (2009) state that principles, norms and certifications aimed at directing corporate actions have emerged as a consequence of the evolution of Corporate Social Responsibility (CSR). At the global level, the pressure to adopt Sustainability Reporting has been given a boost by a number of generally accepted indicators that are championed by certain organisations and special interest groups (Tenuta 2010).

3.1 Corporate Governance and Sustainability Reporting

Corporate governance is anchored on the principle that there is a positive relationship between good governance and compliance with the Law. In South Africa, for example, under the Promotion of Access to Information Act (PAIA), stakeholders have certain rights to company information (RSA 2000). In addition, without the intention to stifle innovation, the King III Report, the standard for corporate governance in South Africa, has adopted an 'apply or explain' and not 'comply or else' approach with regard to disclosure of information (IoD 2009). It stands to reason that the level of reporting detail may vary from one organisation to the next and that sensitive and privileged information should be safeguarded to minimise risks. The disclosed information should be accessible to its intended audience.

Coope (2004) states that corporate reporting material is not easily accessible in some organisations. In fact, some of the organisations that provide reports do so mainly for compliance purposes and the material is not accessible online. The perennial challenge of limited resources has been cited as a major contributing factor for the lack of dynamic online corporate reports. However, the Sustainability Reporting landscape is fast changing in some countries. For example, in the UK and the USA, the publishing of information on social, ethical and environmental risk management is now mandatory (IoD 2009).

In South Africa, companies listed on the Johannesburg Stock Exchange (JSE) are finding disclosure of non-financial information a key to retaining healthy share values (Coope 2004). The traditional report that focused mainly on financial data is proving to be inadequate as information on all aspects of an organisation's life needs to be disclosed comprehensively (Herzig and Godemann 2010; Lozano 2011). However, Eccles (2004) cautions that merely 'ticking governance boxes' will not improve corporate governance. The reporting needs of all stakeholders should be kept in mind in designing organisational reports. A narrow focus on financial aspects for reporting has proven to be inadequate for governance purposes. Increased accountability and transparency demand more than financial reports from organisations. The reporting gaps account for the upsurge in global demand for comprehensive reports on sustainability.

Insenmann et al. (2007), however, warn that Sustainability Reporting has to deal with limitations such as its voluntary status, definition languages, complexity and the emergence of competing frameworks, guidelines and indices. A need exists for organisations across all sectors, including Higher Education, to produce sustainability reports that cover all aspects that are key to their continued existence. This goes a long way in supporting risk identification and management.

The management of risk has also contributed to the growing importance of Sustainability Reporting. Merkel and Litten (2007) state that since sustainability is about balance and risk reduction, HEIs are encouraged to report using financial data (income and expenditure), educational data (degrees granted and research) social data (enrolments) and economic data (impacts). Unfortunately, the focus on environmental data, is lacking and this scenario explains why there could be a growing agitation to include environmental disclosers in sustainability reports. According to Chen and Wongsurawat (2011), holistic organisational reporting can be aided by best practices in Sustainability Reporting such as the GRI G4 template espoused by the Global Reporting Initiative (GRI).

3.2 *Integrated Reporting*

The King III Report underscores the importance of Sustainability Reporting and observes that sustainability should transcend reporting on sustainability and focus on integrated performance. This requires the Board to ensure that the organisation achieves short- and long-term integrated performance goals (IoD 2009).

Integrated reporting, a subset of Sustainability Reporting, encompasses a company's finances and its sustainability and may take the form of one or more reports—all presented at the same time. The integrated report should contextualise the financial report and touch on the achievements and failures in meeting strategic objectives. The report places the responsibility of overseeing integrated reporting on the Audit Committee which should assist the Board with disclosure on sustainability and at times provide assurance on the integrity of the information required (IoD 2009).

Eccles and Armbrester (2011) define integrated reporting as a "holistic and integrated representation of an organisation's performance from a financial and sustainability perspective". Integrated reporting, they add, seeks to give answers to questions such as energy consumption, cost of production, corporate governance and reputational risk, stakeholder satisfaction, service and link to shared values. Eccles and Armbrester (2011) further advise that the International Integrated Reporting Committee (IIRC) was formed in August 2010 with a mandate to develop a globally acceptable framework for accounting for sustainability and that in the same year, 160 companies produced integrated reports using the GRI's G3 guidelines.

In a survey of South African organisations, Sonnenberg and Hamman (2006) show that a limited number of South African companies report according to GRI guidelines although a majority of those surveyed stated their commitment to complying with expectations of the King III Report on corporate governance.

In addition, Sonnenberg and Hamman (2006) further state that South African companies tend to report on sustainability in an aspirational, anecdotal and episodic manner as a result of the lack of regulatory enforcement. This trend implies that many organisations have not fully embraced integrated reporting as a result of various factors such as lack of awareness of the benefits associated with integrated reporting, lack of capacity to report or gaps in the financial systems that generate reporting data.

In an attempt to understand the reasons for limited integrated reporting by organisations, Aras and Crowther (2008) attribute fear of losing competitive advantage to the often cited resistance to full disclosure by companies. Coope (2004) blames limited resources for the lack of dynamic corporate responsibility reporting. Smaller companies are mainly affected. Pennington and Moore (2010) take a broader view and attribute a slow start to factors such as the voluntary nature of Sustainability Reporting, lack of comparability of data across sectors, generalisation of skewed reports, lack of prioritisation of integrated reporting and absence of generally accepted accounting and auditing standards.

Tenuta (2010) avers that the sustainability report is the most operative tool for organisations to communicate with stakeholders. A lack of standards and generally accepted reporting metrics undermine communication. This point is supported by Van den Brink and Van der Woerd (2004) who state that in order to benchmark sustainability performance, there is need for industry-specific benchmarks and formats. The use of prescribed standards and formats will lend more credibility to Sustainability Reporting and allay fears expressed by Lackmann et al. (2012) that most of the sustainability reports are often in qualitative format and therefore of limited use for purposes of financial decision making.

The choice of a reporting approach depends on a number of factors such as availability of data, reporting capacity of an organisation, existing reporting traditions, regulatory requirements and the information needs of the various stakeholders. A Sustainability Report of a Higher Education Institution should, at the very least, cover the economic, social, environmental and educational aspects (Lozano 2006). In order to achieve an integrated sustainability report, organisational data from multiple sources should be collected, processed, analysed and presented in line with the information requirements of stakeholders. Business Intelligence (BI), is a set of tools and technologies that can provide organisations with the capability to produce sustainability reports.

BI refers to the tools an organisation uses to gain a better understanding of operations, markets and competition (Bhatnagar 2009). BI can be viewed as "...a broad category of applications, technologies and processes for gathering, storing, accessing and analysing data to help business users make better decisions" (Watson 2009). BI provides a basis upon which informed decisions can be made in organisations.

4 Research Findings

As part of the survey that was administered, respondents were required to indicate the aspects that their university reports cover. As evident from Fig. 1, most institutions still focus on the economic aspect of reporting.

The results show that environmental reporting is least done at universities. The respondents in the SRPHESA survey indicated that economic (financial 95 %), environmental data (58 %) and 74 % on Social respectively are reported on. This suggests that information from Faculties and Departments on environmental aspects of reporting does not find its way to institutional reports or that it is not mandatory to report on the environmental dimension.

In response to a question on whether specific areas are reported on, the respondents show a lower rate of reporting on Corporate Social Responsibility (CSR) and Environmental Reporting as opposed to reporting on aspects such as financial reporting. Figure 2 shows that almost 50 % of the 23 surveyed (SRPHESA) South African Higher Education Institutions do not report on aspects such as compliance with legislation, impact on the environment and corporate social responsibility or engagement activities.

The main focus of reporting in HEIs is financial reporting as indicated in Figs. 1 and 2. Regulatory requirements and the many tools are available to assist organisations with financial reporting are two possible reasons why these reports take

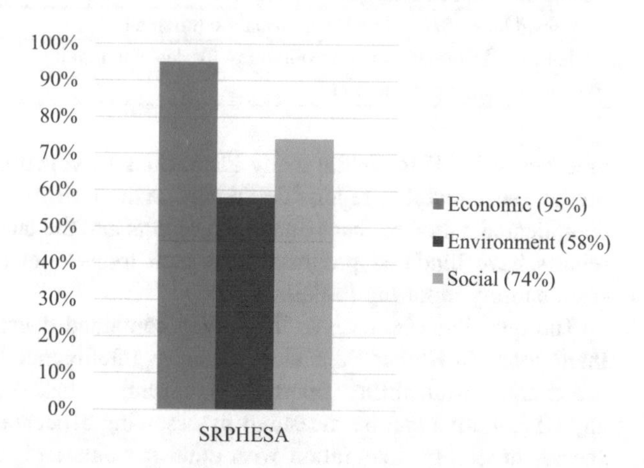

Fig. 1 Dimensions of reporting by universities

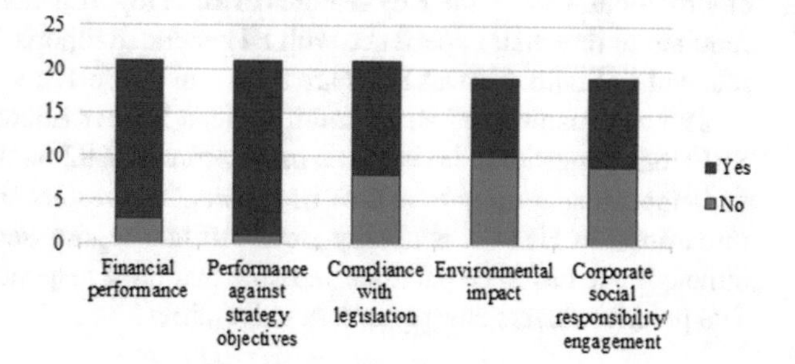

Fig. 2 Aspects covered in reports issued by universities—SRPHESA

Table 1 Experience with reports generated through BI

Experience with BI	SRPHESA (n = 21)	
	Mean	SD
Formats of the reports are pre-determined	2.82	1.29
The frequency of development and distribution of the reports is pre-determined	3.06	0.97
The reports are generated on an ad-hoc basis depending on request	3.47	1.01
BI reports are made available to all relevant users	3.47	0.80
BI reports are availed only to information requesters	3.73	0.70
Users are encouraged and empowered to access BI reports	2.82	1.13

Table 2 Challenges with BI in South African higher education institutions

Challenge	SRPHESA (n = 21)	
	Mean	SD
Unavailability of data	3.62	1.07
Non-existence of data	3.14	1.31
Incompleteness of information	3.62	0.97
Lack of clear information management strategy	3.81	1.08
Lack of integration in reporting systems	3.52	0.93
Limitations with data analytical capability	3.24	0.70
Perceived lack of action on the information provided	3.67	0.97
Staleness of information and unsuitability for decision making	3.10	0.89
Poor information presentation	3.30	0.86

precedence in HEIs. While many institutions surveyed indicated a willingness to produce sustainability reports that report on multiple dimensions of operations, the complicated nature of these integrated reports and the lack of tools to support these reports have hindered progress. One such tool or set of tools that can support sustainability reporting in HEIs is BI.

The questionnaire used in the survey contained questions relating to Business Intelligence in Higher Education. Business Intelligence (BI) tools and techniques can enable sustainability reporting. According to Isik et al. (2013), the success of any BI initiative can be measured by assessing aspects such as data quality, user access, flexibility, integration with other systems and the nature of the decision environment. One of the Key features of BI is report generation. Respondents were required to rate their experience with BI generated reports. The results of the survey relevant to South African HEIs are shown in Table 1.

The low maturity of BI in South African Higher Education is confirmed in the SRPHESA questionnaire with the mean score of 2.82 on the aspect of encouraging and empowering users to access BI reports. The results show that the information turnaround in HEIs is generally slow. Ad hoc reports seem to be most prevalent although one has to be cautious and note that most respondents chose to be neutral. The practice differs sharply across universities.

The low maturity of BI in South African Higher Education could be due to the number of challenges identified by the respondents of the surveys. The challenges of BI in South African HEIs are shown in Table 2.

5 Recommendations and Conclusions

Organisations need information to monitor their performances against set targets. One way of achieving this is through reporting on all facets of an organisation's operations as usually outlined in the strategic plan. Sustainability Reports cover economic, environmental and social aspects of an organisation's existence and therefore provide a platform for tracking performance against its set objectives. Notwithstanding the current reporting practices, HEIs in South Africa need to adopt Sustainability Reporting in line with best practice in governance. Heeks (2006) proposes that public data should be complete, accurate, relevant, timely and appropriate for presentation (CARTA). It takes a carefully planned and iterative process for organisational data to be CARTA. It is envisaged that the introduction and institutionalisation of Sustainability Reporting in HEIs in South Africa will contribute towards achieving CARTA status.

Organisational performance should be reported from a holistic perspective. Stakeholders, including regulatory bodies and the media, are increasingly demanding that organisations disclose information about operations from multiple dimensions as standalone financial reports are no longer seen as an adequate. Organisational reporting has matured in response to the changing requirements of stakeholders. Reporting in South African HEIs is still very narrow in scope with the main focus being on the financial aspect, as legislation requires public HEIs to disclose financial reports. The voluntary nature of reporting on social and environmental aspects has led to minimal reporting and disclosure of this information. The change in the South African Department of Higher Education (DHET) requirements however, requires South African Public HEIs to produce and disclose sustainability reports covering the entire sustainability spectrum from 2015.

In order to produce these integrated sustainability reports, a reporting tool that can integrate data from multiple sources to produce one integrated report will be required. One such tool, or suite of tools, with the required capabilities is BI. BI also provides capabilities to publically disclose sustainability reports to the relevant stakeholders through internet technologies. The use of BI in South African HEIs is still at a low maturity. Disparate reporting systems are currently used in many HEIs, and this presents a challenge in the sustainability reporting process as the final outcome should be one single integrated report. Institutions need to invest in BI technologies that can assist the sustainability reporting process to ensure stakeholder satisfaction and regulatory compliance.

References

Aras G, Crowther D (2008) Developing sustainable reporting standards. J Appl Account Res 9 (1):4–16
Bhatnagar A (2009) Web analytics for business intelligence. Online 33(6):32–35
Chen CH (2011) The major components of corporate social responsibility. J Glob Responsib 2 (1):85–99
Chen CH, Wongsurawat W (2011) Core constructs of corporate social responsibility: a path analysis. Asia Pac J Bus Adm 3(1):47–61
Coope R (2004) Seeing the net potential of online CR communications—learning from the best and worst on the Web. Corp Responsib Manag 1(2):20–25
Dimitrov DK, Davey H (2011) Sustainable development: what is means to CFOs in New Zealand. Asian Rev Account 19(1):86–108
Dumay J, Guthrie J, Farneti F (2010) GRI sustainability reporting guidelines for public and third sector organisations—a critical review. Public Manag Rev 12(4):531–548
Eccles RG (2004) Hopes and fears for financial reporting and corporate governance. Balance Sheet 12(2):8–13
Eccles RG, Armbrester K (2011) Integrated reporting in the cloud. IESE Insight 8:13–20
Fonseca A, MacDonald A, Dandy E, Valenti P (2011) The state of sustainability reporting at Canadian universities. Int J Sustain High Educ 12(1):22–40
Hazelton J, Haigh M (2010) Incorporating sustainability into accounting curricula: lessons learnt from an action research study. Acc Educ 19(12):159–178
Heeks R (2006) Implementing and managing e-government: an international text. Sage, London
Herzig C, Godemann J (2010) Internet-supported sustainability reporting: developments in Germany. Manag Res Rev 33(11):1064–1082
Insenmann R, Bey C, Welter M (2007) Online reporting for sustainability issues. Bus Strateg Environ 16(7):487–501
IoD (2009) King III report on governance for South Africa. Institute of Directors of Southern Africa, Johannesburg
Isik O, Jones MC, Sidorova A (2013) Business intelligence success: the roles of BI capabilities and decision environments. Inf Manag 50(1):13–23
KPMG (2011) KPMG international survey of corporate responsibility reporting. KPMG International, London
Lackmann J, Ernstberger J, Stich M (2012) Market reactions to increased reliability of sustainability information. J Bus Ethics 107(2):111–128
Lozano R (2006) A tool for a graphical assessment of sustainability in universities (GASU). J Clean Prod 14(2):963–972
Lozano R (2011) The state of sustainability reporting on universities. Int J Sustain High Educ 12 (1):67–78
Lozano R, Lukman R, Lozano FJ, Huisingh D, Lambrechts W (2011) Declarations for sustainability in higher education: becoming better leaders, through addressing the university system. J Clean Prod 48:10–19
Merkel J, Litten LH (2007) The sustainability challenge. New Dir Inst Res 2007(134):7–26
Microsoft Dynamics (2010) Environmental sustainability. White paper. Microsoft Corporation, pp 1–14
Pennington LV, Moore E (2010) Sustainability reporting: rhetoric versus reality. Employ Relat Rec 10(1):24–39
Petrini M, Pozzebon M (2009) Managing sustainability with the support of business intelligence: integrating socio-environmental indicators and organisational context. J Strateg Inf Syst 18 (4):178–191
RSA (2000) Promotion of access to information Act (Act No. 2 of 2000). Government Gazette Notice No. 20852. Government Printer, Pretoria
Smith PAC, Scharicz C (2011) The shift needed for sustainability. Learn Organ 18(1):73–86

Sonnenberg D, Hamman R (2006) The JSE socially responsible investment index and the state of sustainability reporting in South Africa. Dev South Afr 23(2):305–320

Stephens JC, Graham AC (2010) Toward an empirical research agenda for sustainability in higher education: exploring the transition management framework. J Clean Prod 18(1):611–618

Tenuta P (2010) The measurement of sustainability. Rev Bus Res 10(2):163–171

Van den Brink JWM, Van der Woerd F (2004) Industry specific sustainability benchmarks: an ECSF pilot bridging corporate sustainability with social responsible investments. J Bus Ethics 55(2):187–203

Watson HJ (2009) Tutorial: business intelligence—past, present and future. Commun Assoc Inf Syst 25(39):487–510

WCED (1987) World commission on environment and development's (The Brundtland Commission) report: our common future. Oxford University Press, Oxford

A Living Lab for Optimising the Health, Socio-economic and Environmental Situation in El Salvador

Melanie Platz, Marlien Herselman, and Jörg Rapp

1 Introduction

One identified critical health problem in El Salvador is chronic renal failure. Nationally, chronic kidney disease is currently the leading cause of hospital mortality in men with a lethality of 11.0 % (MINSAL 2014). Studies in adult population of the Salvadoran farming communities show that the majority of cases with CKD were heavily influenced by factors such as male gender, older age, agricultural occupation, hypertension, family history of hypertension and contact with some pesticides (Orantes et al. 2014).

In order to address this health problem, the project LLinES (2014) proposed to establish a LL in El Salvador to carry out research on low-cost techniques to mitigate exposure to pesticides and other chemicals in the environment, to improve the production processes and to improve the care of kidney patients in rural areas. The project LLinES has the aim to reduce the risk to human health, to local farmers and community members, caused by the exposure to pesticides, through a One Health approach. The One Health approach recognises, that the health of humans, animals and ecosystems are interconnected. Consequently, if risks that originate at the animal-human-ecosystems interface shall be addressed, a coordinated, collaborative, multidisciplinary and cross-sectoral approach is needed (One Health Global Network 2015).

The project LLinES also has the objective to find agricultural productive alternatives without agrochemical inputs or with a rational use of them. By doing so,

M. Platz (✉) • J. Rapp
University of Koblenz-Landau, Landau, Germany
e-mail: platz@uni-landau.de; rapp@uni-landau.de

M. Herselman
University of South Africa & CSIR, Pretoria, South Africa
e-mail: mherselman@csir.co.za

J. Marx Gómez, B. Scholtz (eds.), *Information Technology in Environmental Engineering*, Springer Proceedings in Business and Economics,
DOI 10.1007/978-3-319-25153-0_11

human health and the socio-economic situation are supposed to be optimised, as well as the environmental situation.

The developed risk mitigation strategies and decision support concerning the agricultural production processes shall be provided to the community using Information and Communication Technology (ICT). By analysing the ICT-situation in El Salvador, commonly used information channels shall be identified that enable the distribution of information. If necessary, low cost ICT-technologies to enlarge the ICT-network in El Salvador will be developed, for being able to reach a potentially large target group.

LLinES is mainly focused on mitigating occupational factors and environmental risk for chronic kidney disease (CKD) affecting Salvadoran farmers in endemic and epidemic proportions. The research will be conducted with the involvement of local experts in agriculture, health and environment, as well as the affected population.

The purpose of this paper is to provide an overview of the elements that are essential when establishing a LL and to what extend the establishment of a LL in El Salvador adhered to these elements in order to assist the communities to feel empowered and to address some important social, health and environmental issues in that country. In this context, African LLs are investigated, because the initial situations in Africa and El Salvador are similar relating to the socio-economic and ICT-situation, among others. El Salvador has, as well as Africa particular challenges in relation to rural socio-economic development and sustainable quality of life, due to the current state of available infrastructure, educational and employment opportunities. Consequently, the LL-methods developed in and for Africa might be transferable to LLs in El Salvador.

Therefore, in this paper, core values of LLs will be listed and the purpose of LLs will be explained. Stakeholders of LLs will be addressed and lessons learnt from the LLiSA network will be identified. Afterward, the research methodology will be pictured and a case study will be presented.

2 Core Values of Living Labs

LLs, as a new approach to innovation and ICT development, emerged during the 1990s. It encompasses the idea of creating an environment (e.g. an open innovation environment, also referred to as an ecosystem or platform) that offer users (i.e. different stakeholders such as public-private-partnerships) the opportunity to take part actively in the co-creation of innovation and, more specifically, the development of ICT-related products and services (e.g. idea generation, development, implementation and evaluation) (EnoLL 2014; Følstad 2008; Geerts 2011). A LL can thus be both a milieu (environment, arena) and an approach (methodology, innovation approach).

LLs focus on open innovation, which is a vital element of the knowledge-based economy. This term is coined by Chesbrough (2003) to characterise the fundamental shift from open towards external innovation. Open innovation calls for collaborative organising principles for managing research and innovation. The LL is a way of implementing the open innovation approach (Fahy et al. 2007; Garcia

Guzman et al. 2007; Mulder et al. 2007). From an African perspective the LL can be defined as

> environments, a methodology or an approach which caters for user-driven open innovation within real-life rural and urban settings/communities, where users can collaborate with multiple committed stakeholders (whether NGOs, SMMEs, industrial, academic/research, government institutions or funders) in one or more locations, to become co-creators or co-designers of innovative ideas, processes or products within multidisciplinary environments. Successful deployments can result in improved processes or service delivery, new business models, products or services, and can be replicated (with necessary socio-cultural adaption) to improve overall quality of life and wider socioeconomic impact (including entrepreneurship) in participating and other communities (Cunningham et al. 2011).

This definition can also be applied to any other developing country or context where rural communities are involved. In order to grow LLs in countries which are regarded as developing (like Africa and El Salvador), it is important to ensure wide adoption of this definition and to ensure the involvement from rural communities to strengthen co-creation of innovation and to allow communities to get the direct benefits from all stakeholders involved.

LL environments or platforms have been created for a variety of ICT-related topics from e-commerce to healthcare, transport, tourism development, energy production, agriculture and governance (Smit et al. 2011) as well as specific ICT and IT focus areas like mobile ICT, computing and cognitive systems engineering (Lievens et al. 2006; MacEachren et al. 2006). The importance of LLs in emerging economies has also been indicated in a paper by Smit et al. (2011). Thus in emerging economies contexts, the focus seems to be on the application of ICT-related products and services as catalysts for capacity building and community development or empowerment (Herselman 2011). El Salvador is also regarded as an emerging economy and therefore this is also applicable to them. The core values of a LL can thus be summarised as a milieu and/or an approach that is (Cunningham et al. 2012):

- **Driven by**: Users
- **Involving**: Real-life communities and multiple stakeholders
- **Focused on**: Collaboration and co-creation
- **Aimed at**: Multi-disciplinary open innovation
- **Resulting in**: Improved or new products/processes, replicable and sustainable
- **Impacting on**: The socio-economic environment (capacity building, development, empowerment)

Many of the core values are based in or linked to the users and other stakeholders. The next section expands on stakeholder considerations in the LL environment.

3 Purpose of Living Labs

The importance of LLs as catalysts of innovation is voiced in South Africa and in African countries as developing entities. This has been indicated in the Second Action Plan (2011–2013) of the 8th Africa–EU Strategic Partnership (Science, Information Society, Space). LLs are viewed as a priority area for collaboration between public sector, private sector and research communities in Africa, Europe. And other developing countries like El Salvador. LLs can support science and technology innovation and ICT capacity-building initiatives for mass diffusion of ICTs and related services, as key enablers for poverty reduction, economic growth, social development, optimisation of the health situation and regional integration (Santoro and Conte 2009). The purpose of a LL is therefore threefold:

- Innovation catalyst
- A collaborative environment
- Capacity building

Supporting the establishment and evolution of LLs is seen as a critical tool to enhance ICT research cooperation, local innovation, entrepreneurship and wider socio-economic and community development. LLs are typically established to understand what can ultimately be described as human behavioural responses to ICT. This fits the purpose of the use of LLs in the El Salvador context as LLs are seen as places where inputs from users can be collected and to investigate the context in which ICT is used (i.e. context research), to discover new uses and service opportunities for ICT, to involve users as co-creators and to evaluate new ICT solutions with users in an everyday (user) context (Følstad 2008).

4 Stakeholders of Living Labs

LLs as 'functional places' involve stakeholders from a public private partnership (or the like) of collaborating universities, research institutes, public entities, companies and individuals. The LL approach can assist policy makers to understand what works in practice and what kind of policy environment is needed to sustain the various technology solutions (Santoro and Conte 2009).

Two key dimensions of critical importance, are community engagement—particularly with those traditionally disadvantaged (e.g. women and youth, those living in rural communities), and a greater focus on intangible outcomes (including knowledge and idea creation) within the innovation concept and process.

For the Living Labs in Southern Africa network (LLiSA 2014), a successful LL requires a strategic, mutually beneficial partnership between a minimum of two key stakeholders (e.g. government, industry/business, research/academia, community) with complementary expertise and experience, a common vested interest in the outcomes of enabling users (community) to actively participate in the research, development and innovation process, and at least one stakeholder ensuring the necessary methodological rigor is applied so that results are valid. It is in the social context that one can find various groups interacting with one another and with objects, within a given context (Cunningham et al. 2012).

Challenges which are experienced thus far within rural communities with different stakeholders are to balance the power struggles amongst political structures like rural chiefs or political leaders as well as feelings of entitlement of funders who believe that their rights and needs have to be addressed first as they provide the funding. The top-down and bottom-up support of all parties or stakeholders within the LL is necessary to establish a LL committee to represent all stakeholders and to have the power to make decisions on behalf of the whole community.

Stakeholder considerations for a LL can be summarised as:

- **Types**: Communities, researchers, public entities, private companies, individuals, communities of practice, other LLs
- **Networking**: Sharing knowledge, buy-in, co-creators
- **Knowledge**: Diverse, complementary, including methodological rigour
- **Support**: Top-down and bottom-up
- **Purpose**: Common vested interest

These discussions on the purpose, core values and stakeholders in a LL indicate that the establishment of a LL can be a challenging exercise. The following discussion therefore focuses on lessons learnt and resulting key success factors for the establishment of a LL.

5 Lessons Learnt from LLiSA Network: 2009–2014

LLiSA, which is similar to EnoLL (European network of Living Labs) is a network of LLs which purpose is to create capacity for understanding, establishing and developing LL activities, support pilot projects in Southern Africa and to facilitate local and international collaboration and linkages. It links interested developers, research organisations, industry and government bodies together for advancing regional LL initiatives (Herselman et al. 2009). A key aspect of the LLiSA network is to ensure constant collaboration across LLs as well as a sharing of lessons learnt, contacts with industry partners and recognition (Cunningham et al. 2012; Herselman et al. 2010; LLiSA 2014).

Most of the existing LLs in the LLiSA network have evidence to provide on how they, each with their own uniqueness, are busy with capacity building of communities and with the enhancement of innovation skills with the communities they work with. Most of these LLs are operational in the rural communities scattered throughout South Africa in most of the provinces.

Over a period of 6 years since the establishment of LLiSA a few valuable lessons were learnt, which need consideration when establishing a new LL (Cunningham et al. 2012).

1. The first key success factor is to establish a **commonly owned vision** for the primary goals and objectives through the engagement and a co-creation process with all key stakeholders. While these may vary from country to country and region to region, based on national and regional priorities, educational opportunities, available human and infrastructural resources, urban-suburban-rural divide, transportation and distance challenges and level of socio-economic development, it is also quite likely there will be common goals and objectives.
2. A second key success factor is the need for **strong, focused leadership**, with a clear vision for what the LL aims to achieve, and the support and trust of its members. It is also important that such leadership consists of a group of people with complementary skills, expertise, contacts, credibility and fund raising capacity required to help ensure the LL has the opportunity to achieve its organisational goals. While it might seem obvious that such a leadership group should be representative of its Network members (and this is certainly important to ensure trust and transparency), what might be less obvious (especially as the Network evolves) is that not all the necessary skills, expertise and contacts may be available within the Network itself.
3. A third key factor why LLs within the LLiSA network flourish is that they are **self-sustainable from inception** and do not depend on LLiSA for financial support. Most of the LLs come into existence based on community engagement evidence needed by universities to show impact or from industry for social responsibilities supported or contributed towards. The multi-disciplinary nature of a LL where groups address a centralised problem from different perspectives, but with the same goal works best in testing products and services or for doing research. Industry also finds this as a valuable platform where trust has already been established, as they can test or pilot their proof of concepts with valuable feedback to improve their products and services.
4. A fourth success factor is a strong **sense of community-own challenges** and the ability to be **responsive** to the needs of members, their communities and other stakeholders who either fund LLs or LL Networks or use their services. This should evolve into **community development and skills development** of both its members and their communities, which is critical in a developing country context. This may support cooperation with a local, provincial or national educational institution, government institution, incubator or funding agency with similar or complementary goals.
5. A fifth success factor is that while services can certainly be delivered remotely and virtually (e.g. fast adoption of mobile services, remote training, eHealth, eGovernment services), regular **face-to-face interactions** are important to ensure:

   – Better understanding and management of member expectations
   – Development of trust relationships
   – The overcoming cultural differences; training as necessary
   – Discussion, exploration and development of reward systems that incentivise community loyalty

  - Collaboration opportunities that sometimes can only occur during the "empty spaces" of a meeting or event
  - The regular face-to-face meetings (including annual workshops) facilitated through LLiSA have been identified by members as a key factor that has facilitated progress made to date

6. A sixth key success factor identified from LLiSA's experience, is the potential benefit of hosting or co-locating the LL network with an **existing organisation with strong systems** in place for procurement and payments (essential for operations), and auditing and reporting (which are essential to support third party funding, whether from donor government, national or provincial government, other stakeholders or foundations). Such an organisation should ideally provide access to complementary resources and skill sets useful to the Network and its Members. This hosting organisation could be a thematically relevant government ministry or agency, provincial or national research organisation or Incubator. While selecting such a partner organisation, it is essential their board and senior management are committed to the relationship, there is clear complementarity of purpose (whether reinforcing policy or targeting a common constituency) and mutual benefit.
7. The seventh key success factor addresses the primary types of **support** for a LL to be sustainable. These are funding; capacity building; education/training; knowledge sharing; monitoring and evaluation; content creation; and content licensing.

LLs based Collaborative Open Innovation activities are aligned with key national and sectorial priorities and policies, as this is critical to achieve impact and access public sector financial support. This is why securing Policy Partners as part of the business planning and implementation phase is important. The next section describes an application of these discussions in the establishment of a new LL in El Salvador.

6 Research Methodology

The underlying problem investigated in this research touched on the challenges of establishing a LL that can promote sustainable innovation. The question identified is:

> How is the development of a LL in El Salvador complying with the success factors found in establishment of LLs in South Africa?

This study applied interpretivism as its philosophy and theoretical underpinning. Klein and Myers (1999), define interpretive research as based on knowledge obtained through "social constructivism such as language, consciousness, shared meanings, document tools and other artifacts". The researcher is seen as investigating the phenomenon based on the perceptions of participant's history or experience that the participants have encountered. Walsham (2006) states that the philosophical base of interpretive research is phenomenology and hermeneutics. This is because interpretive research seeks to investigate meanings of words or texts as they are expressed within definite social contexts by various participants according to individuals' previous experiences (Carr and Kemmis 1986). It is in the social context that one can find various groups interacting with one another and with objects within a given context. Therefore interpretive researchers underpin the perceptions of the social actor in order to make sense of the activities that exist within the defined contexts (Hesse-Biber and Leavy 2010).

The Open Community Approach (Niehaus 2013) will be applied on the LL, so that the scientific results of the co-creation, exploration, experimentation and evaluation of innovative socio-economic improvements can be applied in other scenarios or territorial contexts by modifying and adapting the results. If the concepts and related technological artifacts **remain free and open** for other real life use cases, the spreading of successful ideas is more likely. A qualitative approach with a case study strategy is followed, as this allows for an in depth and rich description of a situation (Creswell 2007).

The case study of the design and inception of a LL in El Salvador is described next, to indicate how this LL complies with the elements of a LL and how it will function within the Community to support innovation in addressing environmental and health related issues.

7 Case Study

The success of the LL-concept in Southern Africa gave rise to the establishment of a LL in El Salvador that can act as a hub to foster collaborative, innovative and sustainable research in solving environmental, health and social issues. The LL research approach supports collaborative, open, user-centric and multi-disciplinary research to promote innovative and sustainable solutions for a community's challenges and is therefore an ideal research approach for this unit. The LL provides a formal inter-disciplinary collaborative research space; and access to and formalising of partnerships where community members, government, universities and NGOs as well as external participants can research and seek solutions in the challenging El Salvador environment.

7.1 *The Vision of the El Salvador LL Corresponds to the Threefold Purpose of a Living Lab*

- **Innovation catalyst**: The leading hub for research into innovative solutions for the challenges in the El Salvador environment, recognised for its quality, relevance and impact.
- **Collaborative environment**: An open, user-centric, multidisciplinary research environment, driven by user-communities and their real life experiments, fostering innovation and sustainability in environmental, health and social interventions in El Salvador.
- **Capacity building**: Developing people, creating knowledge and making a difference in the context, environment and social needs of El Salvador locally.
- The core values of the El Salvador LL can be summarised as a milieu and an approach to assist with creating risk awareness for risk mitigation to optimise the socio-economic and health situation in El Salvador.
- **Driven by**: The community Ciudad Romero in El Salvador.
- **Involving**: The prime community consists of researchers at the National University of El Salvador and the University of Koblenz-Landau (Germany), in collaboration with other educators and researchers (teachers, lecturers, industry trainers, pre-service students, post-graduate students and experts in agriculture, health and environment). Other stakeholders are described below.
- **Focused on**: Collaboration and co-creation of innovative solutions for common health and environmental needs and challenges.
- **Aimed at**: Multi-disciplinary open innovation of low-cost techniques to mitigate exposure to pesticides and other chemicals in the environment, to improve in production processes and to optimise the environmental situation and to improve the care of kidney patients in rural areas, incorporating, but not solely focused on ICT.
- **Resulting in**: Improved or new products/processes, replicable and sustainable within each sub-LL focus area.
- **Impacting on**: The socio-economic environment through solutions development and implementation, training modules and workshops.

In the context of the Action Team 6 Follow Up Initiative (AT6FUI 2013), which has the aim to improve public health by the application of space technology, the LL concept is based on a systematic user co-creation approach with farm workers (campesinos) in El Salvador. As an example, the LL research could incorporate the following innovation processes:

- **Technology**: leverage risk mitigation options by using existing technology for a different field of application (e.g. apply a smartphone-app for low cost precision farming, to create an economic and environmental benefit by reduction of the amount of used pesticides and a health benefit by less exposure to agrochemicals).

- **Capacity Building**: create a risk awareness programmes with model experiments for schools. This concept regards children in schools as multipliers of innovation processes.
- **Ecotoxicology**: determine the risk of agrochemicals for the rural population that is suffering from chronic kidney disease and for the environment.
- **OpenSource & Open Content**: existing open source software or open content material for public health risk mitigation can be applied and/or modified in the territorial context (e.g. city, agglomeration, region).
- **Last Mile**: the farm workers might not have direct access to mobile devices, so the user driven innovation in the LL has to consider the last mile between people that have access to ICT and the risk exposed people in rural communities.

The Open Community Approach (AT6FUI 2013) will be applied on the LL, so that the scientific results of the co-creation, exploration and evaluation of innovative socio-economic improvements can be applied in other scenarios or territorial contexts by modifying and adapting the results. If the concepts and related technological artifacts **remain free and open** for other real life use cases, the spreading of successful ideas is more likely. The use cases involve user communities, not only as observed subjects of the LL research, but also as a driver for a sustainable development of creation. For example, farm workers that are exposed to agrochemicals know which problem solving strategies are applicable under the rural working conditions. A useful mask and protective clothes might be unusable under tropical working conditions. Furthermore a proposed solution without consideration of local knowledge could be counterproductive for the objective of risk mitigation, e.g. when the protection increases the dehydration of the body due to heat stress of the body. Therefore, usable solutions have to be developed.

7.2 *Stakeholder Considerations for This LL are Described as Follows*

- **Types**: The Ciudad Romero community, researchers at the National University of El Salvador (academic staff and post-graduate students), other research entities, such as the University of Koblenz-Landau, the Government Departments like the National Institute of Health, Ministry of Health of El Salvador (INS-MINSAL).
- **Networking**: Stakeholders communication and networking are at the core of the LL. This is strengthened by a virtual network, database of members and activities, and organised opportunities for sharing knowledge, buy-in, co-creating.
- **Knowledge**: The nature of the LL is such that it promotes inter-disciplinary collaboration. Each stakeholder's knowledge, expertise and experiences are unique and diverse, but complementary. As the primary community consists of researchers, this knowledge also includes methodological rigour.

- **Support**: The LL was created for the community by the community. Each member brings his own network and expertise. Support is therefor also a network in nature. The institution also provides support on various levels for researchers.
- **Purpose**: The activities of all stakeholders within the centre focus on innovative and sustainable contributions in teaching interventions in order to address the educational challenges in the El Salvador context.

7.3 Key Success Factors as Illustrated in the LL

The last discussion in this section focuses on the key success factors identified by LLiSA, as applied to the sub-LL focus group. The success factors were identified as:

1. Commonly owned vision: Primary goals, institutional memory
2. Strong leadership: Focused, complementary skills
3. Self-sustainable
4. Sense of challenges: community own, nimble and responsive, development
5. Face-to-face meetings
6. Hosted by strong existing organisations
7. Support factors: funding; capacity building; education/training; knowledge sharing; monitoring and evaluation; content creation; and content licensing.

Transferred to the El Salvador context:

1. The commonly owned vision is to create risk awareness for risk mitigation to optimise the socio-economic situation, the health situation and the environmental situation in El Salvador.
2. Strong leadership is ensured by stakeholders like the University of El Salvador, the National Institute of Health of El Salvador and the University of Koblenz-Landau.
3. For the self-sustainability of the LL in El Salvador, sustainable funding has to be acquired for the phase after the launch of the LL. Until the launch of the LL, the project LLinES is funded by the German Federal Ministry of Education and Research.
4. Sense of challenges are community own, as the major community problem of renal failure is supposed to be approached.
5. Face-to-face meetings are enabled in the framework of the LLinES-project.
6. The LL in El Salvador is hosted by strong existing organisations like the University of El Salvador and the National Institute of Health of El Salvador.
7. Support factors: funding should be acquired for the phase after the launch of the LL in El Salvador; capacity building and education/training will be done in the context of creating risk awareness and for improving the production processes

and to improve the care of kidney patients in rural areas and low cost techniques will be developed; in this context, knowledge sharing is crucial.

7.4 Challenges During Development and Inception of the LL

The biggest challenge during the development phase of the LL was to explain the concept of a LL (among others because of the language barrier) and to convince the stakeholders and community members of the concept, as many interventions that failed have already been tried to implement. As soon as the stakeholders and community members understood the concept of a LL, they were very excited and optimistic and stated, that such an approach has never been tried before in their community.

There seemed to be a communication barrier between the local institutions in El Salvador that could be—at least a bit—removed by the meetings initiated by the LLinES project.

It is clear, from the view of this LL, that adhering to the purpose, core values, stakeholder considerations and key success factors as identified by LLiSA were advantageous for the development and inception of the El Salvador LL.

8 Conclusion

Given the considerable contribution provided by LLs to catalysing the Innovation ecosystem, supporting pre-incubation and coaching support for students and emerging enterprises, it is recommended that some public support is available for a grant that can be applied on a competitive basis to support ongoing activities of Innovation Spaces. Funders also have the opportunity to support functions and service development of innovation spaces by influencing new innovative mechanisms to be developed or supported. A core part of the value proposition of a LL, is for Innovation Stakeholders and Network Members to share success stories and lessons learnt, problems encountered and how these were addressed are captured, analysed, generalised and shared. This has immense value to all stakeholders to prevent old mistakes and to innovate even more.

Therefore, in the present paper, the elements of LLs, and how these play a role in the establishment of a LL in El Salvador to assist with creating risk awareness for risk mitigation to optimise the socio-economic, the health situation and the environmental situation were outlined and discussed based on experiences and lessons learnt with LLs in Southern Africa network (LLiSA).

Key success factors identified by LLiSA could be specified for the LL in El Salvador. Consequently, the newly established LL in El Salvador is very promising with collaborations and innovation between communities, industry, academia, learners and schools. This ensured that the LL in El Salvador complies to these success factors and this can influence sustainability in the long term.

The lessons learnt from LLiSA are very useful for the establishment of a LL in El Salvador and they will be considered and, if possible, implemented into the LL in El Salvador to foster the success of the established LL and its future. As a network like LLiSA to ensure constant collaboration across LLs as well as a sharing of lessons learnt, contacts with industry partners and recognition is very valuable, the option to build up such a network in El Salvador in the future could be very helpful.

Acknowledgments We would like to thank the German Federal Ministry of Education and Research who supported the project LLinES, the collaboration and thus the establishment of the LL in El Salvador by a grant. The data presented, the statements made and the views expressed are solely the responsibility of the authors. Furthermore, the work of the project members of the project LLinES was vital for the success of the establishment of the LL. Additionally, we want to thank the National Institute of Health of El Salvador and the University of El Salvador and particularly Rafael Gomez Escoto (University of El Salvador) and Alexandre Ribó (INS-MINSAL) for their support, especially during our visit in El Salvador.

References

AT6FUI (2013) AT6FUI improving public health by application of space technology. http://at6fui.weebly.com/. Accessed 15 June 2015

Carr W, Kemmis S (1986) Becoming critical: education, knowledge and action research. Falmer Press, London

Chesbrough H (2003) Open innovation: the new imperative for creating and profiting from technology. Harvard Business School Press, Harvard

Creswell JW (2007) Qualitative inquiry and research design: choosing among five traditions. Sage, Thousand Oaks

Cunningham P, Herselman M, Cunningham M (2011) Supporting the evolution of sustainable Living Labs and Living Labs networks in Africa. http://www.ist-africa.org/home/files/Supporting_the_Evolution_of_Sustainable_Living_Labs_and_Living_Labs_Networks_in_Africa.pdf. Accessed 15 June 2015

Cunningham P, Cunningham M, Herselman ME (2012) Socio-economic impact of growing Living Labs and Living Lab networks into Africa. In: Proceedings of the IST-Africa 2012 Conference, Dar es Salaam, Tanzania, 9–11 May 2012

EnoLL (2014) Open Living Labs by ENOLL. http://www.openlivinglabs.eu/. Accessed 15 June 2015

Fahy C, Ponce De Leon M, Ståhlbröst A, Schaffers H, Hongisto P (2007) Services of living labs and their networks. In: Cunningham P, Cunningham M (eds) Expanding the knowledge economy: issues, applications, case studies. IOS Press, Amsterdam, pp 713–721

Følstad A (2008) Living labs for innovation and development of information and communication technology: a literature review. Electron J Virtual Organ Netw 10:99–131

Garcia Guzman J, Navarro de la Cruz M, Schaffers H, Kulkki S (2007) Methodological framework for human and user centric rural living labs. In: Cunningham P, Cunningham M (eds) Expanding the knowledge economy: issues, applications, case studies. IOS Press, Amsterdam, pp 730–737

Geerts GL (2011) A design science research methodology and its application to accounting information systems research. Int J Account Inf Syst 12(2):142–151

Herselman ME (2011) Living Labs in Southern Africa network. Paper presented at the 3rd annual LLiSA workshop, Rhodes University, Grahamstown

Herselman ME, Marais MA, Pitse-Boshomane MM, Roux K (2009) Establishing a living lab network in Southern Africa. In: Proceedings of the 3rd international IDIA development informatics conference, Kruger National Park, South Africa, 28–30 Oct 2009

Herselman ME, Marias MA, Pitse-Boshomane MM (2010) Applying Living Lab methodology to enhance skills in innovation. In: Proceedings of the eSkills summit 2010, Cape Town, 26–28 July 2010

Hesse-Biber SN, Leavy P (2010) The practice of qualitative research. Sage, London

Klein HK, Myers MD (1999) A set of principles for conducting and evaluating interpretive field studies in information systems. Manag Inf Syst Q 23(1):67–94

Lievens B, Van den Broeck W, Pierson J (2006) The mobile digital newspaper: embedding the news consumer in technology development by means of living lab research. Paper presented at annual international association for media and communication research conference, Cairo, Egypt, 23–28 July 2006

LLinES (2014) Living Lab en El Salvador. http://llines.weebly.com/the-project.html. Accessed 15 June 2015

LLiSA (2014) Living Labs in Southern Africa. http://llisa.meraka.org.za/index.php/Living_Labs_in_Southern_Africa. Accessed 15 June 2015

MacEachren AM, Cai G, McNeese M, Sharma R, Fuhrmann S (2006) GeoCollaborative crisis management: designing technologies to meet real-world needs. In: Proceedings of the 7th annual international conference on digital government research, San Diego, CA, 21–24 May 2006

MINSAL (2014) Informe de Labores 2013–2014. http://www.salud.gob.sv. Accessed 15 June 2015

Mulder I, Fahy C, Hribernik K, Velthausz D, Feurstein K, Garcia M, Schaffers H, Mirijamdotter A, Ståhlbröst A (2007) Towards harmonized methods and tools for living labs. In: Cunningham P, Cunningham M (eds) Expanding the knowledge economy: issues, applications, case studies. IOS Press, Amsterdam, pp 722–729

Niehaus (2013) Definition: open community (OC). http://at6fui.weebly.com/open-community-approach.html. Accessed 15 June 2015

One Health Global Network (2015) What is one health? http://www.onehealthglobal.net/what-is-one-health/. Accessed 15 June 2015

Orantes CM, Herrera R, Almaguer M, Brizuela EG, Núñez L, Alvarado NP, Fuentes EJ, Bayarre HD, Amaya JC, Calero DJ, Vela XF, Zelaya SM, Granados DV, Orellana P (2014) Epidemiology of chronic kidney disease in adults of Salvadoran agricultural communities. MEDICC Rev 16(2):23–30

Santoro R, Conte M (2009) Living Labs in open innovation functional regions. In: Proceedings of the 15th international conference on concurrent enterprising ICE 2009, Leiden, Netherlands, 22–24 June 2009

Smit D, Herselman M, Eloff JHP, Ngassam E, Venter E, Ntawanga F, Chuang J, Van Greunen D (2011) Formalising living labs to achieve organisational objectives in emerging economies. In: Proceedings of the IST-Africa 2011 conference, Gaborone, Botswana, 11–13 May 2011

Walsham G (2006) Interpreting information systems in organizations. Wiley, Chichester

Modeling the Intention to Use Carbon Footprint Apps

Arno Sagawe, Burkhardt Funk, and Peter Niemeyer

1 Introduction

1.1 Carbon Footprint

In recent years the importance of carbon footprinting (or CO_2 footprinting) has increased from an industry but also an end user perspective. A carbon footprint is a useful mean for determining the environmental impacts of products, services, and other events. With the information of carbon footprinting we can effectively measure and minimize the climate impact. It is an important information that supports climate goals, for example limiting the global warming at 2 °C.

Originally the definition of carbon footprints considered the total sets of greenhouse gas emissions. We follow Wright et al. (2011) and use their definition for carbon footprinting. In their definition, only two prominent greenhouse gases are taken into account: "A measure of the total amount of carbon dioxide (CO_2) and methane (CH_4) emissions of a defined population, system or activity, considering all relevant sources, sinks and storage within the spatial and temporal boundary of the population, system or activity of interest. Calculated as carbon dioxide equivalent (CO_2e) using the relevant 100-year global warming potential (GWP100)."

Carbon footprinting has been discussed in the public (in Germany for example: in education programs and in the media) because it has a much wider audience than other environmental impact categories such as human toxicity or acidification (Pelletier et al. 2007). To follow Weidema et al. (2008): "carbon footprints carry the potential of being a good entry point for increasing consumer awareness and fostering discussions about the environmental impacts of products." At first, in carbon footprinting, things are kept simple, and second, compared to other impact

A. Sagawe (✉) • B. Funk • P. Niemeyer
Universität Lüneburg Leuphana, Lüneburg, Germany
e-mail: sagawe@leuphana.de; funk@leuphana.de; niemeyer@leuphana.de

J. Marx Gómez, B. Scholtz (eds.), *Information Technology in Environmental Engineering*, Springer Proceedings in Business and Economics,
DOI 10.1007/978-3-319-25153-0_12

measures a carbon footprint can be easily estimated (Weidema et al. 2008). Certainly, it shows the consequences with regard to the own carbon footprint when a person realize that 300 cups of filter coffee per year results in a carbon footprint of coarse 0.22 t of CO_2, or more than 2 % of the carbon footprint of an average European in the same time (own calculation).

1.2 Carbon Footprinting Apps

Carbon footprint apps exist for several smartphone operating systems (e.g. Android or iOS), to calculate the personal carbon footprint. Each app has different priorities and features. Some apps capture the CO_2 balance of flights only. Other apps cover all possible data to calculate the personal carbon footprint. Still other apps combine CO_2 balances with fitness programs. For Example, "Carbon Footprint", "Mein CO_2—Fußabdruck", "Carbon Footprint ACP", "CO_2-Footprint", "My Carbon Footprint", "Carbon Emissions Calculator", "Green Footprint Calculator", "CarbonTrack", "Changers—CO_2 fit", to name a few (Table 1). In the maximum a given carbon footprint app has been downloaded only 1000–5000 times (App Store 2015; GooglePlay 2015).

True for most of this kind of apps, the data for calculating a carbon footprint will gather by barcodes. Some apps include challenges, such as: "Reduce your CO_2 consumption in the next 3 days to X percent!", or something similar. Also, these apps allow to share the calculated carbon footprint in social networks.

There are a few research projects that examine how smartphones can be used to support environmental protection and sustainability (Aram et al. 2012). Research on carbon footprint apps does not exist. In 2015 the number of smartphone users around the world is close to 2 bn (Emarketer 2015). Having said that, it is clear, that only a small fraction of smartphone owners has installed or is currently using carbon footprint apps. Knowing that smartphones can have a large impact on user behavior we strongly believe that CO_2 apps could significantly contribute to

Table 1 CO_2 footprint apps

Name	Version	OS	Rating	Votes	Downloads
Carbon Footprint	1.0	Android	3.0/5.0	2	100–500
Mein CO_2—Fußabdruck	1.2	Android	3.5/5.0	7	100–500
Carbon Footprint ACP	1.11	Android	4.8/5.0	5	10–50
CO_2-Footprint	1.0	iOS	4.0/5.0	–	–
My Carbon Footprint	1.0	Android	3.0/5.0	20	1000–5000
Carbon Emissions Calculator	1.1	Android	4.5/5.0	4	1000–5000
Green Footprint Calculator	1.4	Android	1.6/5.0	25	1000–5000
CarbonTrack	1.0	iOS	4.0/5.0	–	–
Changers—CO_2 fit	1.1	iOS	4.0/5.0	–	–

reducing CO_2 emissions caused on an individual level. In order to increase the installed base of CO_2 apps it is important to understand its drivers. Therefore, our research question is "What is the intention to use a carbon footprint app?". We hope to make a contribution that will be considered during the development and publication of carbon footprint apps. In this paper we describe the design and preparation of an empirical study.

2 Research Model and Hypotheses

2.1 Technology Acceptance Model

In order to operationalize the above question, we start from the technology acceptance model (TAM). In information systems research TAM is a model that explains why people use technology or not. It was originally developed by Davis (1989) and published in his thesis. The theoretical underpinnings of the TAM are based on the theory of reasoned action (TRA) (Fishbein and Ajzen 1975). To explain the individual's information technology acceptance behavior, TAM adopts the TRA's causal links (Moon and Kim 2001).

Figure 1 shows the original TAM. External variables, such as an individual's abilities, the type of information technology, the task, and situational constraints, are shown. Davis (1989) defined Perceived Usefulness as: "the degree of which a person believes that using a particular system would enhance his or her job performance" and Perceived Ease of Use as: "the degree of which a person believes that using a particular system would be free of effort" (Moon and Kim 2001). Attitude Toward Using is the degree of which a person intends to use the software. Actual System Usage is the degree of which a person actually use the software. The dashed lines show which constructs are queried by items. The solid lines show the effect of the constructs on each other. Following Davis (1989), these are important factors for evaluating the consequences of using information technology.

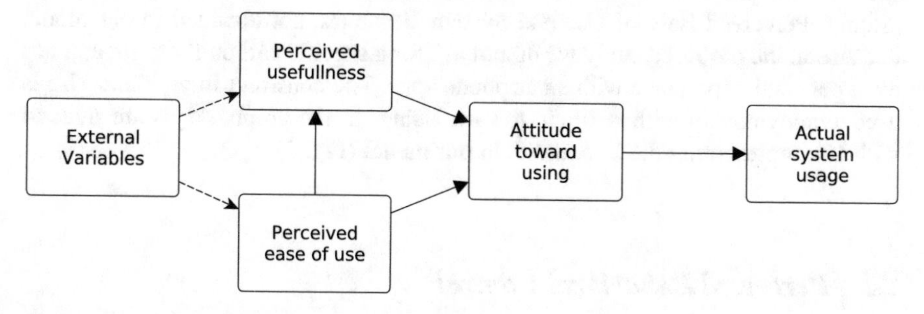

Fig. 1 Technology acceptance model (Davis 1989)

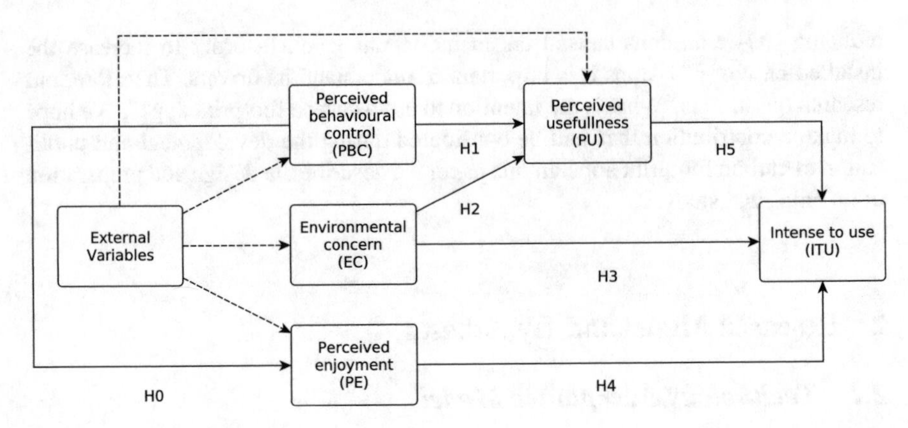

Fig. 2 The research model and hypotheses

TAM was originally developed for working contexts in organizations and companies and, later, adopted to individual settings. For this purpose, additional constructs and external variables were added (Agarwal and Karahanna 2000; Koufaris 2002; Moon and Kim 2001). Reasons for the frequent use of TAM in Information Systems Research may lie in the clarity and simplicity, but also in the high reliability of its input variables, which has been demonstrated in a meta-analysis (King 2006).

Since our context is environmental protection on an individual level, we adapt and extend the original TAM (Fig. 1) by adding the constructs Environmental Concern, Perceived Behavioral Control, and Perceived Enjoyment. We argue that personal Environmental Concerns have a major effect on whether someone will use CO_2 apps. Perceived Behavioral Control relates to two important aspects: first, the technical abilities of a person and, second, the person's view of self effectiveness. Perceived Enjoyment is a well-known construct that captures the playfulness being an important driving force in app usage.

Since socio-demographic variables influence the environmental awareness (Clark et al. 2003; Shen and Saijo 2008; Tanner 1999), we add external variables such as age, place of residence, sex, education, number of kids, and technology affinity. Perceived Ease of Use and System Usage are not included in our model because in the proposed study we do not ask for a specific carbon footprint app nor even personal experience with smartphone apps. The construct Intention to Use is used synonymously with Attitude toward Using. In the graph, edges are marked with Hn, representing the hypothesis in our model (Fig. 2).

2.2 *Perceived Behavioral Control*

For the present study context PBC is related to the consumers' subjective degree of control over adopting and using the carbon footprinting app. We expect, the higher the degree of perceived control the greater the intention to use this kind of app (Kranz and Picot 2011; Lee et al. 2009). PBC is directly linked to perceived usefulness (PU) and perceived enjoyment (PE), as hypothesized. Perceived enjoyment and usefulness were generically found to explain intention to use applications (Verkasalo et al. 2010):

H0: Perceived behavioral control positively influences perceived enjoyment.
H1: Perceived behavioral control positively influences perceived usefulness.

2.3 Environmental Concern

As discussed by Kranz and Picot (2011), environmental psychology defines constructs relating to environmental concern. They are used as predictors of environmental friendly behavior (Kaiser et al. 1999). In accordance with Kranz and Picot (2011) we refer to the construct as environmental concern in this paper. Also we state that environmental concern is positively related to behaviors favoring the preservation of natural resources, i.e. adopting Green-IS (Kranz and Picot 2011).

We expect, people with greater environmental concerns to have a higher likelihood to use a carbon footprint app. Environmental concerns will also increase the perceived usefulness:

H2: Environmental concern positively influences perceived usefulness.
H3: Environmental concern positively influences intention to use.

2.4 Perceived Enjoyment

Perceived enjoyment has a significant effect on the intention to use a new technology (Venkatesh and Bala 2008). Venkatesh and Bala (2008) indicate a positive interaction between perceived enjoyment and the intention to use. This is particularly true for games (Van der Heijden 2004). Smartphones and apps can be used for social or business purposes and for utilitarian as well as hedonic reasons (Van der Heijden 2004). The influence of perceived enjoyment on the acceptance of advanced mobile services (such as apps) is described many times (Hong and Tam 2006; Hong et al. 2006; Kim et al. 2007; Nysveen et al. 2005a, b). Prior studies of mobile data services incorporate perceived enjoyment into the TAM to gain a more accurate prediction of user acceptance toward a specific source (Bruner and Kumar 2005; Moon and Kim 2001; Van der Heijden 2003; Verkasalo et al. 2010).

This study defines perceived enjoyment as the degree to which a person believes that using a carbon footprinting app is interesting and associates adoption with enjoyment. Therefore, we hypothesizes:

H4: Perceived enjoyment positively influences intention to use.

2.5 *Perceived Usefulness*

If an individual perceives an activity to be beneficial to achieve valued outcomes, he or she will be more likely to accept the new technology (Davis 1989). Davis (1989) defines perceived usefulness as the prospective user's subjective probability that using a specific application system will increase his job performance within an organizational context. Adams et al. (1992) follow Davis and find perceived usefulness to be a major determinant of use behavior and intention. Several studies examine the correlation between perceived usefulness and the intention to use in evaluating consumer acceptance of an innovative product (Moon and Kim 2001; Van der Heijden 2003; Vijayasarathy 2004).

We define perceived usefulness as the degree to which an individual perceives the use of a carbon footprint app helps to be more efficiency in living an environmental lifestyle:

H5: Perceived usefulness positively influences intention to use.

2.6 *Intention to Use*

Finally, in our model, the intention to use is positively influenced by environmental concerns, perceived enjoyment, and perceived usefulness. According Davis and Olson (1985) the intention to use is positively influenced by perceived ease of use. As discussed above, in our TAM the perceived ease of use is not included.

3 Study Design

For the proposed study we use an online questionnaire based assessment. First, the participants of the survey will be presented a fictitious carbon footprint app and its main features are described (see Sect. 4.1). With the fictitious app in mind, study participants are asked questions that provide insights into the intention to use a carbon footprint app. This enables us to target the general population and not just a small group of people which already uses this type of app.

Our questionnaire includes three types of questions: first, control questions (Table 2), second, socio-demographic questions (Table 3) and third, the main

Table 2 Control questions

Scale items	Question	Value
S	Do you have a smartphone?	[Yes v no; 0 v 1]
C1	Do you know what is meant by the term carbon footprint?	[Yes v no; 0 v 1]
C2	If so, what is your simple definition of it?	[Varchar; 2048]

Table 3 Variables describing characteristics of participants

Scale items	Name	Value
A	Age	18–n
P	Place of residence	{GER, EU, Earth}
S	Sex	{Male, female}
G	Graduation	{Primary school, secondary school, college, university degree, PhD, other}
W	Work	{Employed, Pensioner, Scholar, Student, House-wife, Other}
N	Number of kids	[0–x]
T	Technology affinity	[no-absolute; 0–6]
S	Smartphone use	[never-permanent; 0–6]

questions relating to the constructs used in our study. The control questions are used to identify relevant participants that are characterized by having a smartphone and know about the concept behind carbon footprinting. The constructs Perceived Behavioral Control (PBC), Environmental Concern (EC), Perceived Enjoyment (PE) und Perceived Usefulness (PU) are questioned with five items per construct (Table 4). Likewise, the Intention to Use (ITU) is questioned.

We acquire study participants via crowd sourcing. In these kind of settings, participants are often called Clickworkers, a term, originally coined by NASA (Haythornthwaite 2009). Today, working with Clickworkers is an emerging trends in data collection (Gosling and Mason 2015). Clickworkers are internet users who subscribe to a crowd sourcing provider (CSP) in order to respond to small tasks on a fee basis. Amazon Mechanical Turk and the Germany based company Clickworker. de are two examples for CSPs, sometimes also called micro job sites. To ensure the quality of work, Clickworkers are tested by the CSP regarding their qualification. To do so Clickworkers are required to provide their skills, knowledge and interests during the registration process. Some CSPs offer online exams to new Clickworkers. With respect to our study there are no special requirements to the participants apart from the above control questions. But, could it be that a participants is interested only in the money? Then the participants would answer the survey as soon as possible. Therefore, we use the time as an indication of the seriousness of the answers. If the time is less than 3 min, the survey cannot be evaluated.

Table 4 Research constructs and operational definitions

Scale items	Constructs	Sources
Perceived behavioral control (PBC)		
PBC1	1. I can use the app without help from others	Taylor and Todd (1995), Verkasalo et al. (2010)
PBC2	2. Using the app is entirely within my control	
PBC3	3. I have the resources and the knowledge and the ability to make use of the app	
PBC4	4. My knowledge of carbon footprint is sufficient to use the app	
PBC5	5. With the app I have control over my carbon footprint	
Environmental concern (EC)		
EC1	1. The current civilisation is destroying nature	Kranz and Picot (2011), Fraj and Martinez (2006), Haanpää (2007), Dunlap and Van Liere (2008)
EC2	2. If the necessary measures are not taken the environment deterioration will be irreversible	
EC3	3. Whenever possible, I travel by train instead of flying in order to protect the environment	
EC4	4. I throw garbage in selective containers	
EC5	5. Humans have the right to modify the natural environment to suit their needs	
Perceived enjoyment (PE)		
PE1	1. The app must be interesting.	Davis (1989), Moon and Kim (2001), Van der Heijden (2004), Liao et al. (2008), Nysveen et al. (2005a), Bruner and Kumar (2005), Igbaria et al. (1994)
PE2	2. Document my carbon footprint makes me feel enjoyable.	
PE3	3. Using the app is a good way to spend my leisure time.	
PE4	4. A variety of functionality in app arouse my curiosity.	
PE5	5. Using the app involves me in the enjoyable process	
Perceived usefulness (PU)		
PU1	1. The app makes my life easier as an environmentally conscious person	Davis (1989, 1992), Adams et al. (1992), Subramanian (1994)
PU2	2. Using the app facilitates the efficacy of my ecological lifestyle	
PU3	3. Using the app supports my lifestyle	
PU4	4. Using the app increases the quality of my life	
PU5	5. Using the app is useful for me	

(continued)

Table 4 (continued)

Scale items	Constructs	Sources
Intention to use (ITU)		
ITU1	1. If I have access to the app, I would use it	Luarn and Lin (2005), Moon and Kim (2001), Jackson et al. (1997)
ITU2	2. I intend to using the app in the future	
ITU3	3. I would recommend others to use the app	
ITU4	4. I use such an app already	
ITU5	5. I will not use such an app	

The study targets the general population in German speaking countries. We decided to not go for a more international setting because we would expect a high bias towards specific countries. Men and women will be questioned ($n \geq 200$). The participants must be 18 years old, a requirement enforced by the CSP we use for our study.

4 Conclusion and Future Work

In this article, we described our upcoming study and the underlying Research Model and Hypotheses to measure the intention to use Carbon Footprinting Apps. Therefore, we customized the original TAM and adapt respectively extend the original model by adding constructs which addresses environmental protection on an individual level: Environmental Concern, Perceived Behavioral Control, and Perceived Enjoyment. These constructs have been defined in further works from other authors who deal with similar topics.

Our study will be carried out in the second quarter of 2015. The digital questionnaire will be provided on our web servers and the collected data will be stored on our database servers. Participants receive a link to the questionnaire. Only those participants that complete the questionnaire will receive their fee via the micro job site. Estimating the structural equation model will be done by using smartPLS which implements the partial least square approach. First results will be presented during the ITEE conference.

Appendix

App Description

Suppose you would have access to the following app on your smartphone: The app under consideration allows you to monitor your personal carbon footprint that occurs when consuming products and energy intensive services. The app provider is an EU based company that takes serious measures to protect your data. The download and use of the app is free. Overall, the app is self-explanatory, given the fact that you understand the concept of carbon footprints.

The data used to calculate the carbon footprint can be easily collected. For data collection multiple ways are possible: the smartphone camera is used as a scanner. In addition the app reliably and intelligently interprets sensor data and action logs (e.g. calendar entries) on the phone to infer the carbon footprint from traveling. If products are equipped with NFC tags touching the product is sufficient to determine the carbon footprint and give you the choice to make part of your carbon footprint, e.g. when buying the product. The app has extensive and fun-to-use reporting facility that help you to understand your daily and monthly carbon footprint and set it into the context of climate change. Thus you can grasp what your contribution to the problem is. In addition, the app gives recommendations how to act differently ("take your bike today and save . . .") and supports competitions on social media platforms like Facebook ("who is the most environment friendly person in your network today?").

References

Adams DA, Nelson RR, Todd PA (1992) Perceived usefulness, ease of use, and usage of information technology: a replication. MIS Q 16(2):227–247

Agarwal R, Karahanna E (2000) Time flies when you're having fun: cognitive absorption and beliefs about information technology usage. MIS Q 24(4):665–694

App Store (2015) Apps for iOS smartphones. http://store.apple.com/com. Accessed 15 Feb 2015

Aram S, Troiano A, Pasero E (2012) Environment sensing using smartphone. In: ITEE sensors applications symposium 2012, University of Brescia, Brescia, 7–9 Feb 2012

Bruner GC, Kumar A (2005) Explaining consumer acceptance of handheld Internet devices. J Bus Res 58(5):553–558

Clark CF, Kotchen MJ, Moore MR (2003) Internal and external influences on pro-environmental behavior: participation in a green electricity program. J Environ Psychol 23(3):237–246

Davis FD (1989) Perceived usefulness, perceived ease of use, and user acceptance of information technology. MIS Q 13(3):319–339

Davis FD (1992) Extrinsic and intrinsic motivation to use computers in the workplace. J Appl Soc Psychol 22(14):1111–1132

Davis GB, Olson MH (1985) Management information systems: conceptual foundations, structure, and development, 2nd edn. McGraw-Hill, New York

Dunlap RE, Van Liere KD (2008) The "new environmental paradigm". J Environ Educ 40 (1):19–28

Emarketer (2015) Worldwide smartphone usage to grow 25% in 2014. http://www.emarketer.com/Article/Worldwide-Smartphone-Usage-Grow-25-2014/1010920. Accessed 29 Apr 2015

Fishbein M, Ajzen I (1975) Belief, attitude, intention and behavior: an introduction to theory and research. Addison-Wesley, Reading, MA

Fraj E, Martinez E (2006) Environmental values and lifestyles as determining factors of ecological consumer behaviour: an empirical analysis. J Consum Mark 23(3):133–144

GooglePlay (2015) Apps for Android smartphones. https://play.google.com/store. Accessed 15 Feb 2015

Gosling SD, Mason W (2015) Internet research in psychology. Annu Rev Psychol 66:877–902

Haanpää L (2007) Consumers' green commitment: indication of a postmodern lifestyle? Int J Consum Stud 31(5):478–486

Haythornthwaite C (2009) Crowds and communities: light and heavyweight models of peer production. In: Proceedings of the 42nd annual Hawaii international conference on system sciences, Big Island, 5–8 Jan 2009

Hong S-J, Tam KY (2006) Understanding the adoption of multipurpose information appliances: the case of mobile data services. Inf Syst Res 17(2):162–179

Hong SH, Tam K, Kim J (2006) Mobile data service fuels the desire for uniqueness. Commun ACM 49(9):89–95

Igbaria M, Schiffman SJ, Wieckowski TJ (1994) The respective roles of perceived usefulness and perceived fun in the acceptance of microcomputer technology. Behav Inform Technol 13 (6):349–361

Jackson CM, Chow S, Leitch RA (1997) Toward an understanding of the behavioral intention to use an information system. Decis Sci 28(2):357–389

Kaiser FG, Wölfing S, Fuhrer U (1999) Environmental attitude and ecological behaviour. J Environ Psychol 19(1):1–19

Kim H-W, Chan HC, Gupta S (2007) Value-based adoption of mobile internet: an empirical investigation. Decis Support Syst 43(1):111–126

King WR (2006) A meta-analysis of the technology acceptance model. Inf Manag 43(6):740–755

Koufaris M (2002) Applying the technology acceptance model and flow theory to online consumer behavior. Inf Syst Res 13(2):205–223

Kranz J, Picot A (2011) Why are consumers going green? The role of environmental concerns in private Green-IS adaption. In: Proceedings of the 19th European conference on information systems (ECIS) at AIS Electronic Library (AISeL), Helsinki, Finland, 9–11 June 2011

Lee H-J, Lim H, Jolly LD, Lee J (2009) Consumer lifestyles and adoption of high-technology products: a case of South Korea. J Int Consum Mark 21(2):153–167

Liao C-H, Tsou C-W, Shu Y-C (2008) The roles of perceived enjoyment and price perception in determining acceptance of multimedia-on-demand. Int J Bus Inf 3(1):27–52

Luarn P, Lin H-H (2005) Toward an understanding of the behavioral intention to use mobile banking. Comput Hum Behav 21(6):873–891

Moon JW, Kim YG (2001) Extending the TAM for a World-Wide-Web context. Inf Manag 38 (4):217–230

Nysveen H, Pedersen PE, Thorbjørnsen H (2005a) Intentions to use mobile services: antecedents and cross-service comparisons. J Acad Mark Sci 33(3):330–346

Nysveen H, Pedersen PE, Thorbjørnsen H (2005b) Explaining intention to use mobile chat services: moderating effects of gender. J Consum Mark 22(5):247–256

Pelletier NL, Ayer NW, Tyedmers PH, Kruse SA, Flysjo A, Robillard G, Ziegler F, Scholz AJ, Sonesson U (2007) Impact categories for life cycle assessment research of seafood production systems: review and prospectus. Int J Life Cycle Assess 12(6):414–421

Shen J, Saijo T (2008) Reexamining the relations between socio-demographic characteristics and individual environmental concern: evidence from Shanghai data. J Environ Psychol 28 (1):42–50

Subramanian GH (1994) A replication of perceived usefulness and perceived ease of use measurement. Decis Sci 25(5–6):863–874

Tanner C (1999) Constraints on environmental behaviour. J Environ Psychol 19(2):145–157

Taylor S, Todd PA (1995) Understanding information technology usage: a test of competing models. Inf Syst Res 6(2):144–176

Van der Heijden H (2003) Factors influencing the usage of websites: the case of a generic portal in The Netherlands. Inf Manag 40(6):541–549

Van der Heijden H (2004) User acceptance of hedonic information systems. MIS Q 28(4):695–704

Venkatesh V, Bala H (2008) Technology acceptance model 3 and a research agenda on interventions. Decis Sci 39(2):273–315

Verkasalo H, López-Nicolás C, Molina-Castillo FJ, Bouwman H (2010) Analysis of users and non-users of smartphone applications. Telematics Inform 27(3):242–255

Vijayasarathy LR (2004) Predicting consumer intentions to use on-line shopping: the case for an augmented technology acceptance model. Inf Manag 41(6):747–762
Weidema BP, Thrane M, Christensen P, Schmidt J, Løkke S (2008) Carbon footprint. J Ind Ecol 12 (1):3–6
Wright LA, Kemp S, Williams I (2011) 'Carbon footprinting': towards a universally accepted definition. Carbon Manag 2(1):61–72

Mass Customization: Sustainability of a Computer-Based Manufacturing System

Hans-Knud Arndt

1 Introduction

The issue of mass customization becomes increasingly meaningful to the global economy. Whether cars or cereals, shampoos or sneakers: not only small start-ups but also global companies such as Adidas and Coca-Cola follow this trend. Is this only hype or actually a business model for the future? The second question is about the sustainability of mass customization. The third question concerns the role of the information and communication technology (ICT) with both the realization of mass customization and the realization of a sustainable mass customization.

2 Mass Customization

Mass customization is a concept introduced by Stanley M. Davis in 1987. Traditional industrial production is using economies of scale to produce goods at low costs. This is called mass production. Mass production has nearly no variety (for example the Ford T Model). In contrast, individual production with a high degree of variety has small volumes with high costs. Davis suggested that new business models should accept "the coexistence of mutually contradictory phenomena" (Davis 1989) and he called this business model "mass customization". Mass customization is realizable through the use of new technology that offers both economies of scale and specificity. The goal is to produce goods that are individually and cost-effectively manufactured. Besides questions of manufacturing, the customer's

H.-K. Arndt (✉)
Otto-von-Guericke-Universität Magdeburg, Magdeburg, Germany
e-mail: hans-knud.arndt@iti.cs.uni-magdeburg.de

J. Marx Gómez, B. Scholtz (eds.), *Information Technology in Environmental Engineering*, Springer Proceedings in Business and Economics,
DOI 10.1007/978-3-319-25153-0_13

awareness and perception of mass customization is mentioned as another prerequisite of the concept (Davis 1989).

A very common definition of mass customization is that it refers to the production of products and services for a (relatively) large market, which meets the different needs of each customer of these products, with the efficiency of a comparable mass or series production (Piller 2006). According to this definition, the following four aspects of mass customization should be considered (Piller 2004):

- **Customer Co-Design**: The genus of mass customization is customer co-design (Piller 2004). Customers are integrated into the product and service creation by defining and configuring an individual solution. It is a concretization of the end user desires and needs to certain product specifications. For this procedure, the end user needs a tool, e.g., a paper-based catalog or configuration software (Boër et al. 2013).
- **The Needs of Each Individual Customer**: The co-design procedure provides all possible product configurations offered by the manufacturer (degree of customization). The goal of mass customization is a single customized product, which correctly identifies the customization dimensions and options meant to satisfy the customer needs. These individual needs can be operationalized by the following three dimensions (Boër et al. 2013):
  - *Fit*: measurements of a product with the given dimensions of the recipient
  - *Style*: aesthetic design by modifications aiming at optical or sensual senses
  - *Functionality*: technical attributes (power, interfaces etc.) of an offering.

A three-branch radar graph depicts these dimensions of customization (see Fig. 1).

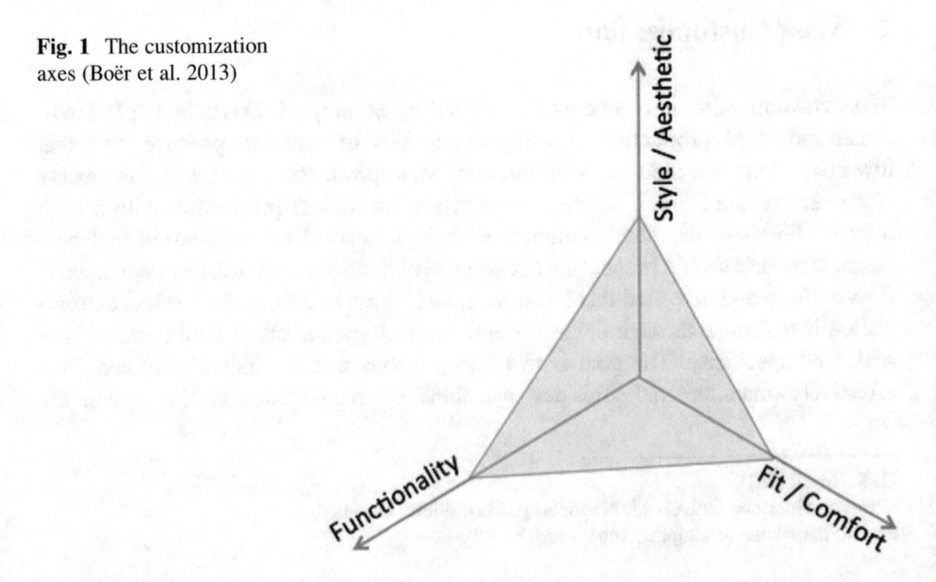

Fig. 1 The customization axes (Boër et al. 2013)

- **Stable Solution Space**: A solution space can be defined as pre-existing capability and degrees of freedom built into a given manufacturer's production system (Von Hippel 2001). Therefore, a successful mass customization system is characterized by *stable* but still flexible and responsive processes that provide a dynamic flow of products (Piller 2004). While within a conventional (craft) customization the customizer re-invents not only its products but also its processes for each individual customer, a mass customizer uses stable processes to provide high-variety products and services.
- **Adequate Price**: Mass customization can be distinguished from craft customization by the fact that mass-customized products and services are targeting the same market segment like the markets of corresponding standard products and services. In contrast, the traditional craft customization is located in the premium price segment. Therefore, the targets of craft customization are completely different market segments.

In the field of environmental sustainable mass customization research, one basic question is whether "mass-customized products [are] more sustainable or less sustainable compared to traditionally mass produced products" (Brunø et al. 2013). Methods used for this kind of research question usually focus on the difference of modular products (mass customization) and integrated products (traditional mass production). Initially, it is assumed that integrated products have the advantage that "the performance of a product can be improved compared to a modular product" (Brunø et al. 2013).

Research results (Brunø et al. 2013) point out that there are several relations between the elements of mass customization and environmental sustainability. Typical concepts, which have an impact on the performance of mass customization, are:

- Individual distribution
- Modularity
- Process variety
- Product variety
- Reconfigurability.

All these concepts have an effect on the following indicated elements of environmental sustainability in the context of mass customization:

- Energy efficiency
- Material usage
- Process variety
- Need for additional products
- Reuse
- Remanufacturing
- Enables upgrades
- Service
- Premature disposal

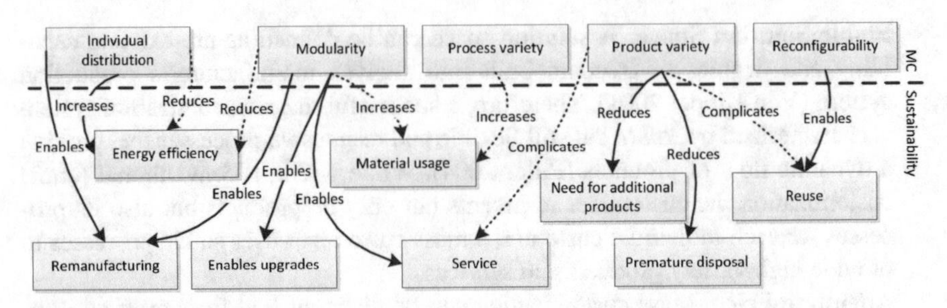

Fig. 2 Relations between mass customization and sustainability (Brunø et al. 2013)

Concerning the element of energy efficiency, the concept of individual distribution potentially has a positive influence (solid line in Fig. 2) and a negative influence (dotted line in Fig. 2) on the environmental sustainability compared to mass production. Furthermore, individual distribution enables remanufacturing (positive effect on environmental sustainability). The concept of modularity has negative effects on the energy efficiency and material usage (both dotted lines in Fig. 2) and positive effects on remanufacturing, enables upgrades and service (solid lines in Fig. 2). The concept of process variety increases the material usage (negative effect, dotted line in Fig. 2). The concept of product variety complicates the elements service and reuse (negative effects, dotted lines in Fig. 2) and reduces the need for additional products as well as the premature disposal (positive effects, solid lines in Fig. 2). Finally, the concept of reconfigurability enables reuse (positive effect, solid line in Fig. 2).

This qualitative study concludes that eight positive and six negative relationships were identified. However, these numbers cannot be used for concluding that mass customization is more sustainable than mass production, since these relations are not unambiguously quantifiable and can only be quantified for specific products as different products have different environmental impacts. Furthermore, the authors of the study state that further research could analyze these relations using a quantitative approach for specific product types (Brunø et al. 2013).

The qualitative study of Brunø et al. (2013) points out positive and negative relations between mass customization and sustainability, but provides no starting point for an approach of sustainable mass customization. Thinking about (more) sustainability in the context of mass customization, the main connecting factor is the aspect *customer co-design* in conjunction with the aspect *needs of each individual customer*.

3 Customer Co-design as Connecting Factor of Sustainability

The customer is of particular importance to mass customization. The basis of the value-added process is the co-design process for the definition of individual goods and services in interaction between providers and users (Piller 2006). Therefore, the customers must be involved in the product design process so that a specific product is produced that meets their needs and requirements.

The basic idea is that the customer co-design process is performed based on a list of options and predefined components. This list is the result of surveys and analyses (aspect *needs of each individual customer*) before the customization process. “Those options were defined trying to meet the needs of the individual customer by analyzing the needs of the many.” (Boër et al. 2013). Keeping the thesis “identifying the needs of the individual customer by analyzing the needs of the many” guides us to one fundamental question of design, especially in the context of sustainability:

> Is analyzing the needs of the many able to identify the best and sustainable solution space for mass customization?

A wrong solution space leads us to what the former chief designer of Braun GmbH, Dieter Rams, calls “the arbitrariness and the thoughtlessness with which many products are designed today” (Hein 2015) such that resources are wasted. In his “ten principles for good design” (Vitsoe 2015) Rams states that:

- **Good design is long-lasting** (principle 7): “It avoids being fashionable and therefore never appears antiquated. Unlike fashionable design, it lasts many years—even in today’s throwaway society”.
- **Good design is environmentally friendly** (principle 9): “Design makes an important contribution to the preservation of the environment. It conserves resources and minimizes physical and visual pollution throughout the lifecycle of the product.”

Therefore, the question of how to implement the customer co-design process of mass customization towards sustainability is also a question of how to achieve good design. More precisely: Is the mass able to generate individual and sustainable customer demands or would it be better if the designer creates appropriate options for individual customers? Furthermore, are the customization axes style/aesthetic, fit/comfort and functionality of the mass customization aspect “the needs of each individual customer” the right dimensions for a sustainable mass customization? In detail: Is a single customer able to enumerate in advance their needs of functionality, style/aesthetic and comfort, e.g., for a new car like the new Volkswagen Golf model, in the distant future? The personal fit belonging to the individual bodily space circumstances is the exception. For example, the mass-customized fit of a sport shoe is certainly an interesting factor. But the decision of shoe color and shoe pattern is not a question of personal fit. In order to not displease the latest fashion trends, the customer needs certain design proposals from the shoe designers.

Design is always to be seen in the relationship of a certain technology and appropriate markets and is never to be seen as a scope in itself (like artists do). The Grand Management Information Design is an approach that deals with these three aspects of design, technology and markets towards sustainable solutions.

4 Grand Management Information Design

Basically every production of goods and services is imaginable as mass customization. If and when a specific goods or service can actually be realized as mass customization depends on the realized or realizable combination of design, technology and market. In 2005, the company Braun GmbH, which belongs to the US company Procter & Gamble, described this approach as "Grand Design", which (at that time for physical products) raises the aspects of design and technology so that new markets can be created and developed, respectively.

A good example of this approach is Apple's iPad (which is in line with Braun's design tradition) that was introduced by Steve Jobs in 2010. Lots of individual manufacturers (including Apple) started several attempts in the field of tablet computers with first experiments in design, technology and market penetration of tablet computers. Though, it is the special combination of design, technology and market response to the iPad, which succeeded.

Our Grand Management Information Design approach is based on the idea of Braun's Grand Design. The goal of Grand Management Information Design is to support the users of ICT with sustainably designed equipment according to their sustainable needs (see Fig. 3). Apple's iPad underlines the purposes of Grand Management Information Design (Arndt 2014):

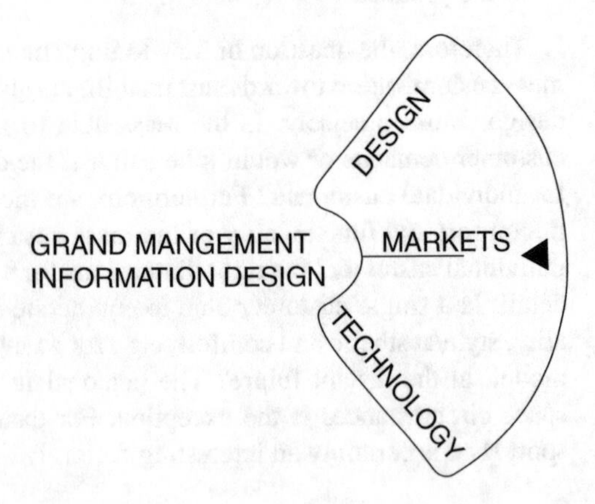

Fig. 3 Grand Management Information Design

- **Design**: The technical quality of the iPad significantly expresses the elegance of both hardware and user interface (software).
- **Technology**: The iPad has significantly increased the technical quality of mobile end devices with touch screens.
- **Market**: Despite initial skepticism, the iPad has succeeded in creating a new market of utmost importance.

In order to achieve an environmental and sustainable **design** of ICT ecosystems, we first have to look at the design process of Apple. While Apple traditionally focuses on ideas and intuition of a visionary designer, in the past the competitor Microsoft developed new products and services by analyzing the—hardly clearly articulated—demands of many and by testing usability in large scale (Arndt 2013). The former Microsoft design approach is very similar to the *needs of the many approach* stated in the context of mass production (see above). According to the ten principles of Rams, good design is the key to greater environmental friendliness and sustainability of products. This also applies to ICT products and ICT ecosystems. The chance to achieve a continuously good design is much higher using the Apple design approach than using the former Microsoft approach.

The question of **technology** can also be explained by one of the ten principles of good design of Rams. In principle 1 “Good design is innovative” (Vitsoe 2015), Rams states: “The possibilities for innovation are not, by any means, exhausted. Technological development is always offering new opportunities for innovative design. But innovative design always develops in tandem with innovative technology, and can never be an end in itself.”

The iPad exemplifies the question of **market**. When the iPad was introduced in 2010, the people asked if a device like an iPad is really essential. Is there a significant added value by using such a device? Does the iPad have specific functions that differ from other tablet computers? (Arndt et al. 2013). In an interview with the French Online Service Buzz Média Orange-Le Figaro a Swiss publisher said that the iPad is not more than dalliance (Haymarket Media 2010). Media journalist Jeff Jarvis explained why he returned his iPad to the store (Höly 2010): “I’m taking my iPad back to the store [. . .]. It’s really because I don’t see a need for it. It’s solving a problem that I don’t think exists [. . .].” In contrast, in 2010, a few months after the release of the first iPad, Steve Jobs predicted that tablet computers would eventually overtake Personal Computers (PCs). “In 2014, according to updated figures from Gartner, after five years of rather crazy tablet growth and slowly declining PC sales, 2015 will be the year that Jobs’ post-PC dream is finally realized. In 2015, Gartner predicts a total of 320 million tablet sales versus only 316 million PC sales (desktops and laptops)” (Anthony 2014). Steve Jobs created a new and very important market in the ICT area with tablet computers.

5 Mass Customization: A Computer-Based Manufacturing System and a Manufacturing System for Information and Communication Technology

Modern business and industrial concepts can only be realized if suitable information and communication technology is available and new ICT leads to new opportunities in business and industry. Therefore, the idea of mass customization can only be achieved as a computer-based manufacturing system, and ICT products can also be manufactured as mass-customized products.

Today, Apple's iPad is still not a product which is attributable to mass customization. But we can see that Apple makes significant steps towards diversification. The iPhone is now available in two sizes (iPhone 6 and the larger iPhone 6S) and more than the original two color versions (three for iPhone 6 and 6S, and even five colors for iPhone 5C). Diversification leads to mass customization. However, at which point exactly this step is taken away from the industrial mass production towards mass customization is only decided by the respective realizable or realized combination of design, technology and market.

The Internet plays a crucial role in mass customization. More generally, it may not only be the Internet, but a corresponding information and communication platform. On the customer side, product configurations can be realized with the help of the Internet, i.e., realizing the front end of mass customization, and the Internet transports the user interface to the customer. In the background, the communication along the value chain or rather value network of customized products is based on the Internet as well (supply chain management).

New ICT will always lead to new opportunities for products and services. For example, the 3D printer will lead to acceleration effects in mass customization, but the exact amount is difficult to predict. Certainly, time savings, such as for the development of new products, are possible. And of course, the 3D printer will lead to new applications of mass customization. A mass customization is always limited. But these limits are not fixed in the course of time, they are continuously changing.

Certainly, there are specific limits of extra charges caused by mass customization. Today, it is difficult to imagine that products, which you can buy in five-and-dimes, were produced in terms of mass customization. But who says that this fact will be valid forever—particularly for companies that are competing in our globalized economy? Especially with 3D printers it is possible to think about an implementation of mass customization by printing cheap products on-site in five-and-dimes in the future. However, this vision may cause large negative environmental impacts. Therefore, design, technology and market must be considered in a synopsis to achieve successful and sustainable ICT products and ICT-based solutions such as mass customization.

6 Mass Customization and Sustainability

ICT represents a key technology that helps to offer continuously new and ever more individualized services on the market. Mass customization leads to diversification and vice versa. Diversification—at least in a company—is limited. In this context, the keyword *core competencies* is important. Companies have only limited core competencies available in a certain time. The designer Dieter Rams states that leaving the company's core competencies leads to arbitrariness, which is dangerous for the long-term success in the market. Furthermore, arbitrariness is contrary to the sustainable and respectful use of resources and causes negative environmental impacts.

However, the inexactly determinable boundary between industrial mass production and individualized mass customization has to be approximately identified in each period. This boundary is influenced by the kind of product (e.g. automotive industry) or service. Premium products are rather subject to mass customization than volume products (especially observable in the automotive sector). Each time has a price-based limit for mass customization, but with a certain combination of design, technology and market there is and will be a move away from premium products to volume products. One the other hand, the opportunities of customization of premium products will always be far greater than the opportunities of customization of volume products.

A particular manner of handling the question of sustainability and mass customization is an approach which addresses the link between mass customization and sustainability by defining a set of key performance indicators (KPIs) such as the sustainability assessment model (SAM) (Boër et al. 2013) or the Mass Customization Business Model (MC BM): The "development of an MC BM provides the required backbone to analyze the environmental impacts of such a BM. In other words, we can monitor the performance of the MC BM in terms of environmental sustainability to understand 'Is mass customization is sustainable?'. In order to track the possible impacts, three KPI are defined in this study: waste production, energy consumption and emission" (Pourabdollahian et al. 2014). More generally, "providing customers with enough information on the environmental impact of certain product attributes during the co-design process can help the users to understand the sustainability impact of their individual choices and thus ultimately supports the design of more ecofriendly products" (Pourabdollahian and Steiner 2014).

Though, we must state that the simple addition of KPIs to the concept of mass customization is far from being a sustainable mass customization. In order to achieve a sustainable mass customization, we need an operationalization of the Grand Management Information Design approach. A core point of the Grand Management Information Design approach and the concept of mass customization is the **design** of products and services. Design is always to be seen in the relationship of a certain technology and appropriate markets, operationalized by the ten principles of good design of Dieter Rams (Vitsoe 2015):

1. **Good design is innovative:** The possibilities for innovation are not, by any means, exhausted. Technological development is always offering new opportunities for innovative design. But innovative design always develops in tandem with innovative technology, and can never be an end in itself.
2. **Good design makes a product useful:** A product is bought to be used. It has to satisfy certain criteria, not only functional, but also psychological and aesthetic. Good design emphasizes the usefulness of a product whilst disregarding anything that could possibly detract from it.
3. **Good design is aesthetic:** The aesthetic quality of a product is integral to its usefulness because products we use every day affect our person and our well-being. But only well-executed objects can be beautiful.
4. **Good design makes a product understandable:** It clarifies the product's structure. Better still, it can make the product talk. At best, it is self-explanatory.
5. **Good design is unobtrusive:** Products fulfilling a purpose are like tools. They are neither decorative objects nor works of art. Their design should therefore be both neutral and restrained, to leave room for the user's self-expression.
6. **Good design is honest:** It does not make a product more innovative, powerful or valuable than it really is. It does not attempt to manipulate the consumer with promises that cannot be kept.
7. **Good design is long-lasting:** It avoids being fashionable and therefore never appears antiquated. Unlike fashionable design, it lasts many years—even in today's throwaway society.
8. **Good design is thought-out to the last detail:** Nothing must be arbitrary or left to chance. Care and accuracy in the design process show respect towards the user.
9. **Good design is environmentally friendly:** Design makes an important contribution to the preservation of the environment. It conserves resources and minimizes physical and visual pollution throughout the lifecycle of the product.
10. **Good design is as little design as possible:** Less, but better—because it concentrates on the essential aspects, and the products are not burdened with non-essentials. Back to purity, back to simplicity.

Therefore, these ten principles of good design should consequently be applied to the concept of mass customization. And of course, during the design process and all other processes of mass customization sustainable KPIs are needed. But it must be kept in mind that some of these KPIs are measurable and some are not. The Global Reporting Initiative (GRI) guidelines for sustainable reporting provide a good overview of potential sustainability performance indicators, which are organized by economic, environmental and social categories (GRI 2011):

- Economic Performance Indicator:

– Aspect: Economic Performance (4 indicators)
– Aspect: Market Presence (3 indicators)
– Aspect: Indirect Economic Impacts (2 indicators)

- Environmental Performance Indicators:
  - Aspect: Materials (2 indicators)
  - Aspect: Energy (5 indicators)
  - Aspect: Water (3 indicators)
  - Aspect: Biodiversity (5 indicators)
  - Aspect: Emissions, Effluents, and Waste (10 indicators)
  - Aspect: Products and Services (2 indicators)
  - Aspect: Compliance (1 indicator)
  - Aspect: Transport (1 indicator)
  - Aspect: Overall (1 indicator)
- Social Performance Indicators:
  (a) Labor Practices and Decent Work Performance Indicators:
    - Aspect: Employment (4 indicators)
    - Aspect: Labor/Management Relations (2 indicators)
    - Aspect: Occupational Health and Safety (4 indicators)
    - Aspect: Training and Education (3 indicators)
    - Aspect: Diversity and Equal Opportunity (1 indicator)
    - Aspect: Equal Remuneration For Women And Men (1 indicator)

  (b) Human Rights Performance Indicators:
    - Aspect: Investment and Procurement Practices (3 indicators)
    - Aspect: Non-discrimination (1 indicator)
    - Aspect: Freedom of Association and Collective Bargaining (1 indicator)
    - Aspect: Child Labor (1 indicator)
    - Aspect: Forced and Compulsory Labor
      Core (1 indicator)
    - Aspect: Security Practices (1 indicator)
    - Aspect: Indigenous Rights (1 indicator)
    - Aspect: Assessment (1 indicator)
    - Aspect: Remediation (1 indicator)

  (c) Society Performance Indicators:
    - Aspect: Local Communities (3 indicators)
    - Aspect: Corruption (3 indicators)
    - Aspect: Public Policy (2 indicators)
    - Aspect: Anti-Competitive Behavior (1 indicator)
    - Aspect: Compliance (1 indicator)

  (d) Product Responsibility Performance Indicators:
    - Aspect: Customer Health and Safety (2 indicators)
    - Aspect: Product and Service Labeling (3 indicators)
    - Aspect: Marketing Communications (2 indicators)
    - Aspect: Customer Privacy (1 indicator)
    - Aspect: Compliance (1 indicator)

7 Conclusion

The goal of Grand Management Information Design is to support the concept of mass customization with sustainable products and services according to the sustainable needs of end users. Therefore, in the context of mass customization the implementation of customer co-design is always to be seen as a sustainable pre-design of an expert in the relationship of a certain technology and appropriate markets. KPIs are not the key to an approach of sustainable mass customization. After all, we have highlighted that KPIs in this respect are merely a supplement of design—but a reasonable one.

References

Anthony S (2014) In 2015 tablet sales will finally surpass PCs, fulfilling Steve Jobs' post-PC prophecy. http://www.extremetech.com/computing/185937-in-2015-tablet-sales-will-finally-surpass-pcs-fulfilling-steve-jobs-post-pc-prophecy. Accessed 22 Mar 2015

Arndt H-K (2013) Umweltinformatik und Design—Eine relevante Fragestellung? In: Horbach M (ed) INFORMATIK 2013: Informatik angepasst an Mensch, Organisation und Umwelt. 43. Jahrestagung der Gesellschaft für Informatik, Koblenz, Sept 2013. Gesellschaft für Informatik, vol P-220, Bonner Köllen Verlag, pp 931–939

Arndt H-K (2014) Big data oder grand management information design? In: Plödereder E, Grunske L, Schneider E, Ull D (eds) INFORMATIK 2014: Big data—Komplexität meistern. 44. Jahrestagung der Gesellschaft für Informatik, Stuttgart, Sept 2014. Gesellschaft für Informatik, vol P-232, pp 1947–1956

Arndt H-K, Mokosch M, Pleshkanovska R (2013) iPad—an environmental-friendly working tool? In: Page B, Fleischer AG, Göbel J, Wohlgemuth V (eds) EnviroInfo 2013: environmental informatics and renewable energies. 27th conference on environmental informatics—informatics for environmental protection, sustainable development and risk management, Hamburg, Sept 2013. Shaker Verlag, Aachen, pp 492–502

Boër CR, Pedrazzoli P, Bettoni A, Sorlini M (2013) Mass customization and sustainability: an assessment framework and industrial implementation. Springer, London

Brunø TD, Nielsen K, Taps SB, Jørgensen KA (2013) Sustainability evaluation of mass customization. In: Prabhu V, Taisch M, Kiritsis D (eds) Advances in production management systems: sustainable production and service supply chains. IFIP WG 5.7 international conference, APMS 2013, State College, Sept 2013. IFIP advances in information and communication technology, vol 414. Springer, Heidelberg, pp 175–182

Davis SM (1989) From "future perfect": mass customizing. Strateg Leadersh 17(2):16–21

GRI (2011) Sustainability reporting guidelines. https://www.globalreporting.org/resourcelibrary/G3.1-Guidelines-Incl-Technical-rotocol.pdf. Accessed 22 Mar 2015

Haymarket Media (2010) Auch Michael Ringier hält den Ball flach: "Das iPad ist bislang nur eine Spielerei". http://kress.de/mail/tagesdienst/detail/beitrag/104288-auch-michael-ringier-haelt-denball-flach-das-ipad-ist-bislang-nur-eine-spielerei.html. Accessed 22 Mar 2015

Hein B (2015) Design guru Dieter Rams says Apple built his dream PC. http://www.cultofmac.com/316222/design-guru-dieter-rams-says-apple-built-his-dream-pc/. Accessed 22 Mar 2015

Höly D (2010) Steve Jobs und das iPad: Kein Messias und nur eine Maschine (3). https://juiced.de/6005/steve-jobs-und-das-ipad-kein-messias-und-nur-eine-maschine-3/. Accessed 22 Mar 2015

Piller FT (2004) Mass customization: reflections on the state of the concept. Int J Flex Manuf Syst 16(4):313–334

Piller FT (2006) Mass Customization: Ein wettbewerbsstrategisches Konzept im Informationszeitalter. Deutscher Universitätsverlag, Wiesbaden

Pourabdollahian G, Steiner F (2014) Environmental and social impacts of mass customization: an analysis of beginning-of-life phases. In: Grabot B, Vallespir B, Samuel G, Bouras A, Kiritsis D (eds) Advances in production management systems. Innovative and knowledge-based production management in a global-local world. IFIP WG 5.7 International Conference, APMS 2014, Ajaccio, Sept 2014. IFIP advances in information and communication technology, vol 439. Springer, Heidelberg, pp 526–532

Pourabdollahian G, Taisch M, Piller, FT (2014) Is sustainable mass customization an oxymoron? An empirical study to analyze the environmental impacts of a MC business model. In: Brunoe TD, Nielsen K, Joergensen KA, Taps SB (eds) Proceedings of the 7th world conference on mass customization, personalization, and co-creation, Aalborg, Feb 2014. Lecture notes in production engineering. Springer, pp 301–310

Vitsoe (2015) Dieter Rams: ten principles for good design. https://www.vitsoe.com/gb/about/good-design. Accessed 22 Mar 2015

Von Hippel E (2001) PERSPECTIVE: user toolkits for innovation. J Prod Innov Manag 18(4): 247–257

Printed in the United States
By Bookmasters